75¢

The Food and Hea[...]eenhouse

Yanda

Fisher

Library of Congress Catalog Card No. 79-91276
ISBN 0-912528-20-6

Published by
John Muir Publications, Inc.
P.O. Box 613
Santa Fe, New Mexico 87501

Sixth Printing, Revised and Expanded

Printed in the United States of America

TABLE OF CONTENTS

ACKNOWLEDGEMENTS

This book spans six years of work and hundreds of solar greenhouses built, observed and fondled. The final result couldn't have occurred with the efforts of only two individuals. Here are some of the people who helped make it happen:

All the contributors to Chapter VIII
The innovators and individualists, thousands of them, across the U.S. who aren't afraid to try something new.
Keith Haggard
Peter van Dresser
Joan Loitz
Frances Tyson
Fran Wessling
Jack Park
B.T. Rogers
The Balcombs and the Nichols
Jene Lyon for permission to reprint sections and drawings from *An Attached Solar Greenhouse*
Robert Fisher
Claude and Lynn Phipps
The entire Robert Bunker Family
Elmer and Hazel Yanda
The New Mexico Solar Energy Association

Special thanks to Ken and Barbara Luboff and J. Stick for their support and hard work in making the book a reality.

More thanks to Andrew McGruer for photo printing work.

Finally, the two most important to us:

Lisa Gray Fisher whose bountiful greenhouse garden contributed to the results described in Chapter VII. Lisa also provided indispensable direction and skill in editing both editions of this book.

Susan Bunker Yanda for thousands of miles covered and just about as many solar greenhouses built and created. Her overall knowledge in managing and teaching are evident in every page of this volume.

Together, Lisa and Susan wrote Chapter VII, the Greenhouse Garden.

In Chapters I through VII the illustrations were done by Rick Fisher and the photographs by the authors unless otherwise noted.

INTRODUCTION

First, a definition is in order because there is some confusion created by the term "solar greenhouse." The confusion is understandable because all greenhouses are, in fact, solar. However, traditionally designed greenhouses have rarely been concerned with the most effective use of the sun's energy. Those described in this book are. We have incorporated four basic elements in the design and operation of each of our greenhouses:

1. The most efficient collection of solar energy.
2. The storage of solar energy.
3. The reduction of heat loss during and following collection periods.
4. Zone layout for the particular light and temperature requirements of various plants.

Attention to these elements produces the following benefits:

1. Surplus thermal energy produced in winter can be used immediately in an adjoining structure or stored for later use.
2. Independence from mechanical heating and cooling devices powered by fossil fuel.
3. Fresh food and colorful flowers right through the winter.

This book, the designs and the benefits derived from it, all come from a basic concern with people's relationship to their environment. One basic environmental problem is centered around *misuse* of energy. We realized that while many people wish for alternative systems, the success of such systems is dependent on the individual's commitment to the system coupled with an understanding of what makes it work. And we want you to know exactly what's involved in building and maintaining your own solar unit.

In the following pages, we've shown methods that can be used to make an appreciable addition to the quality of your life through a closer involvement with your food chain (fresher and cheaper vegetables), a free source of partial heating for your house, a more realistic integration with the cycles of the sun, the seasons and the weather, and independence from corporate energy and food games. Whether or not you actually build a greenhouse depends on many factors: space, economics, appropriateness to your location, and determination, to name a few. But even if you don't build, reading this book will enlarge your understanding of your environment and your relationship with it.

This book grew out of the Solar Sustenance Project begun in 1974. It was a modest demonstration project to determine if attached greenhouses could supplement homes in eleven high-elevation locations in the Rockies with fresh food and heat throughout long and cold winters. The work has evolved into an educational process that has worldwide relevance. The solar greenhouse is unique in that it can satisfy two basic human needs, food and shelter. With other beneficial side effects, such as water conservation and distillation, the potential for greenhouse application is just beginning to be understood.

When we began the project, many engineers and architects insisted that our simple greenhouses wouldn't lengthen the growing season even a week. We were told by others that the 90-degree heat produced by the units was virtually useless. Fortunately, we didn't listen to them. Balancing the negativism of the cynics, we had the support of many people in the field: Keith Haggard and Peter Van Dresser of Santa Fe, T.A. Lawand of the Brace Institute in Quebec, Dr. Francis Wessling of the University of New Mexico, and several of the people mentioned in Chapter VIII. Now, competent professionals from all over the world are eagerly exploring the solar greenhouse field, and their expertise will certainly advance the state of the art.

An important aspect of solar greenhouses is that the principles of design can be applied at any economic level. The $7.00 recycled lumber and polyethylene greenhouse slapped to the south side of a dilapidated dwelling can be just as important and valid a solar application as a $200,000 new solar greenhome under construction in a nearby resort community. We've tried to show the whole range of greenhouses in this book and let you make the decision about where you want to jump in.

Our work on the project and on this book is founded on two principles: the first is that food production should be a low-energy process. The process is begun by growing as much as you can at home, avoiding anything requiring more units of energy to produce than it contains. For that reason, highly controlled, close tolerance food production techniques relying on outside energy sources to maintain them are not stressed here.

The second principle is that greenhouses and other habitable structures should be designed to make maximum use of natural energy flow and to make minimum use of fossil fuels. This means designing a "passive" structure with proper orientation, thermal mass and good insulation. This is not a new idea, but it is being re-examined today in the light of present technological capabilities. While a passive structure delivers obvious benefits, it also demands more thought, design work, labor and care in building.

In many ways the passively designed structure is in direct opposition to the current American mode of living. It's not temporary by nature. The structure itself has a "thermal momentum" that is much like the physiological processes of a human body, charging and discharging, inhaling and exhaling. Most importantly, a well-designed passive structure doesn't depend on a constant supply of energy to keep it livable. The building uses the sun as the earth does, only better.

Since our initial work was done, thousands of passive solar greenhouses have been built. Recent computer simulation studies and advanced technical reports have shown these to be feasible in *any* climate that has heating needs and some winter sunlight to capture and store. We'd like to think that the first edition of our book did a great deal to stimulate scientific interest in and examination of the potential of solar greenhouses. We know from experience that this book has been widely applied. It is a user's book; you who buy it will most likely be building a greenhouse or solar application shortly. Depending on where you live, you may need to increase the performance of your unit through modifications in design or addition of more sophisticated heat collection and storage systems. For those to whom this applies, we've presented a wide range of such improvements in Chapters IV, V and VIII.

The solar greenhouse field has been blessed with many innovators who are also superb teachers and lecturers. The calm confidence of Doug Balcomb, the lucidity of Susan Nichols, the aesthetic impeccability of David Wright, the humor and directness of Jack Park, the patient explanations of Doug Taft, and the work of many other talented individuals has done far more to promote solar usage than any government feasibility studies or private advertising campaigns. People like these have contributed immeasurably to the growing use of solar energy.

When you decide to build and operate a solar greenhouse of your own, you will be joining a group of experimenters in what is still an infant science. You do not need to be a scientist to participate. All the principles involved are elementary and logical. Their simplicity makes the benefits derived from becoming an active member of the solar community easily accessible to you. Welcome.

CHAPTER I THE GREENHOUSE BIOSPHERE

The concepts of *environment* and *ecosystem* have been around for a long time, but only in the past few years have these ideas become part of the public awareness. Many of us only realized the profound implications of these concepts when we saw the first photographs of the earth taken from space by the astronauts. The earth is indeed a closed system, one that must sustain itself through a harmonious balance of its elements.

When you build your greenhouse, you will be creating a very special space, an earth in microcosm. You will control the character of the space to a great extent. Your imagination and design will determine how well the natural life force sustains itself and what you derive from it in return. You are, in effect, producing a *living* place that will grow and evolve with a life force of its own.

The special environment that you will create is a *biosphere*. Webster's definition of a biosphere is: "A part of the world in which life can exist. . .living beings being together with their environment." As a living being, you are an essential element in maintaining your biosphere. Sowing seeds, nurturing the earth, watering, fertilizing plants and soil, and controlling the temperature and humidity will be your contribution to the biosphere. The greenhouse will reward you with the personal fulfillment of living within the cycle of growth.

Figure 1

Solar greenhouses vary greatly in the number of their components and life systems, depending upon the interest, time, and energy invested in them. A simple, easily maintained example would consist of a small structure with a few planting areas. Closely related, hardy varieties of vegetables and/or flowers would be chosen for cultivation. As their needs are similar, they would not require a great deal of time or attention. You may, however, prefer maintaining a complex unit containing a variety of life forms. Some experimental greenhouses of this type combine plant growth (soil or nutri-culture) with the production of

Figure 2

animal protein in the form of fish and rabbits. These systems attempt to achieve a symbiotic balance between the various organisms, using the by-products and waste of each to support the other. The more complex environments may also employ wind generators to power independent heat collectors, sophisticated storage facilities and other improvements (Chapter VIII). These systems obviously demand much more time, attention and strong interest in experimentation.

As a living space, your greenhouse will grow and affect things around it. If it is attached to your house or another structure, an interaction between the two will occur. The conditions that develop in the greenhouse will be shared with an adjacent room or building in the forms of heat, humidity, and the exhilarating fragrance of growth. In addition to pure sensuous delight, there can be economic benefits through a reduction in heating costs and food bills. The changing moods of the life system will soon become evident, and you may find yourself reacting to them much as you would to a human personality.

Along with the rewards are the health benefits that you will enjoy. Greenhouse-fresh produce, especially if it is organically grown, can be far superior to its supermarket counterpart. Commercially produced foods may contain harmful chemicals, and in many cases lose much of their food value during the days they are in transit and on the shelf. Not only will your body welcome the added nutrition of home-grown produce, but you will also experience an unbelievable increase in flavor from the fresh vegetables. The environment of the greenhouse can also produce a feeling of well-being and tranquility. It may become a spiritual refuge from the outside world.

Perhaps the most dynamic aspect of your newly created biosphere is its relationship to the life force outside of our earth's environment—the sun. Solar energy affects every facet of life and change on earth. The sun produces movement in the atmosphere, water and land masses. It acts upon the earth's orbit and seasonal changes. Its waves of visible and invisible energy are the basis of all growth and life. This awesome force will be the medium through which you work. You will collect its energy, contain and store it, alter and direct it in the way most beneficial to the support of your biosphere. The sun will combine with air, earth and water to produce the fifth essential element in the greenhouse, plant life. In the management of your greenhouse, your role will be to complete this five-sided cycle.

CHAPTER II THE DEPENDENCE CYCLE

The mass-market age is the mass-dependence age. Dangerous aspects of the dependence cycle are self-evident. Dependence is addiction. Whether it's a dock loader's strike in Philadelphia or a twenty-cent jump in the per-gallon price of gasoline, the result is the same. Changes are made in your life, usually for the worse, without your having any say in the matter. Urbanization is part of this cycle; specialization in employment is as well. Everyone in this country has felt the effects of this situation and suffered some of the consequences. When those consequences affect basic life functions, it becomes a serious problem. The question is, "How do *you* break the dependence cycle?"

Going back to the land is one method, but for the majority of people, those who live and work in urban areas, this isn't a viable alternative. Rural life isn't everyone's dream and it's difficult, to say the least, to turn a 40' x 80' city lot into a self-sufficient farm. But one doesn't need to be *entirely* dependent on the system. A greenhouse makes it possible to grow a substantial amount of food in a very small area. Moreover, it lengthens the growing season tremendously in most parts of the country and protects crops from damage by hail, wind, and animals.

In order to prevent trading dependence on one part of the cycle for another, a basic rule of thumb is to make a careful examination of how much energy goes into food production from seed to table, then compare that with the amount of energy that comes out of the food to an animal or person. Think about how much energy it takes to grow, harvest, pack, store, and ship the lettuce in your salad and you'll quickly see what that means. Consider gasoline and oil for tractors and trucks, energy expended to drill that oil, to transport roughnecks to the oil fields, to generate the electricity used in supermarket freezers and lighting, and so on. And on. It adds up. Obviously a thoughtful long-range food/energy view takes production techniques into consideration, giving top priority to "low-energy-in, high-energy-out" approaches.

Again we come back to the family or community-operated greenhouse. It's hard to find a better example. It shortcuts the entire process. The family that grows a head of lettuce realizes a measurable petrochemical savings. Shipping costs are eliminated. Food is eaten fresh from the earth; no processing or packaging costs are involved. And it is produced by human labor without machine (purchase, operation, and maintenance) expenses.

Aside from economic benefits, the pleasure of raising your own fresh, flavorful food ecologically and a feeling of self-reliance are additional rewards.

For all the above reasons, private greenhouse sales have increased tremendously. But the problem with buying prefabricated greenhouses or plans is that they were designed without regard for the specific climate and solar conditions in your region, and they weren't planned for your site or your house. In fact, the majority of prefab greenhouses are designed as freestanding structures which demand additional fossil fuel in winter. Rather than adding heat to your home, they actually increase your consumption of fuel.

While we obviously haven't been able to see your home or your site, we've provided enough basics along with design modifications and information on how to use them, that you'll be able to use this book, save some money, understand why your greenhouse is working, and best of all, end up with a life support system custom designed for your home.

CHAPTER III

PRINCIPLES

The principles involved in the dynamics of a solar greenhouse are shared by all solar applications. Here are some of the factors that apply specifically to the solar greenhouse.

Solar Radiation. Energy from the sun strikes the earth constantly and is called *radiation* or *insolation*. It is in the form of *direct, diffuse,* and *reflected* rays. Direct radiation occurs in clear sky conditions. Diffuse is caused by cloud cover, atmospheric conditions, or manmade conditions such as smog. Reflected radiation is bounced from objects, snow, water, clouds, or the ground itself.

The two major components of solar radiation will both be used in the greenhouse. The visible range is used by the plants for *photosynthesis:* the conversion of light, carbon dioxide, and water into food for the plants. *Thermal* or *infrared* radiation is heat. It is created when the visible light strikes objects inside the greenhouse.

Light Collection

Light Energy for Plants. The amount of time a plant receives light determines the amount of food it can manufacture. The *photoperiod* is the relative lengths of light and darkness and their effect on plant development. Plants fall into three categories in terms of their light requirements: short day or winter (a few flowering plants), long day or summer (fruiting vegetables), and neutral or year-round producers (leafy greens). Factors such as location of plants in the greenhouse, their arrangement, and the number and placement of reflective interior wall surfaces are important for promoting good plant growth. Plant growth rate is determined by the *intensity* of light and the length of time light is available. Different plants require different intensities of light, but usually photosynthesis occurs adequately at one quarter of the maximum potential light intensity. The greenhouse designs in this book have enough clear surface to provide sufficient light for photosynthesis.

Percent of Possible Sunshine. The amount of sunshine that reaches the ground in a particular place is expressed as a percentage of the total amount that is possible in a year. The following list gives this information for five major cities in the United States:

Albuquerque 76% Denver 67% Chicago 59% New York 59% Seattle 45%

In planning a solar greenhouse, a knowledge of monthly or seasonal trends is as important as the annual solar percentage. For instance, mid-Michigan has a pattern of extremely cloudy weather from October through December. In January, although the temperatures are colder than in the fall, the solar conditions improve greatly and supplementary heating from an attached greenhouse is more readily available than in October. Monthly or daily technical data on solar availability is most valuable when supported by personal experience. (For seasonal percentage of sunshine see p. 180.)

Solar Collection. Because the glazing of the greenhouse traps a certain amount of the sun's energy, we can think of the greenhouse as a solar collector. It is a solar collector for itself and also for the structure it adjoins. How much solar energy it collects at various times of the year and under different weather conditions is dependent on many factors. Building orientation is one of the important ones.

The majority of the clear glazing in a solar greenhouse *must* face a *southerly* direction, because in the northern hemisphere the sun is in the southern sky throughout the cold winter months.

Because the sun spends the winter in the south, that is the direction from which most of the solar energy is coming. (Of course, the earth is orbiting around the sun, and the tilt of its axis accounts for the change in seasons, so the sun doesn't really go south in the winter. The position of the sun, as we describe it, is actually *apparent* movement from a fixed location on earth.) By facing south, the greenhouse is able to capture the maximum amount of winter sunlight.

The following chart compares solar transmission through east and west, southeast and southwest, and due south-facing vertical glass walls. The amount of solar energy coming through one square foot of glass is given in B.T.U.s (British Thermal Units). One B.T.U. is the heat energy required to raise one pound of water one degree Fahrenheit. For now, let's say that you need hundreds of thousands of them daily in order to have substantial heat for your home.

As you can see from this chart, due east and west clear surfaces are very poor winter collectors, but excellent for solar gain in summer (as you have probably noticed if you have a large west-facing window in your home). However, surfaces that face as much as 45° to the east or west of south receive approximately two-thirds of the winter direct sunlight of south-facing vertical glass. This gives you a great deal of flexibility in design (see **Greenhouse Configurations**, p. 25).

SOLAR TRANSMISSION THROUGH VERTICAL DOUBLE GLASS AT THREE ORIENTATIONS

Ground Reflection Assumed at .2
In BTU/square foot per day

Orientation to South	90° to East or West				45° to S. East or S. West				0° South			
	Dec.	June	Mar.	Sept.	Dec.	June	Mar.	Sept.	Dec.	June	Mar.	Sept.
Latitude North												
Latitude 36°	463	1056	882	842	1083	818	1153	1100	1527	446	1146	1102
Latitude 40°	393	1083	858	816	1007	883	1183	1126	1435	527	1243	1191
Latitude 44°	307	1116	829	787	895	968	1206	1144	1292	628	1324	1263
Latitude 48°	237	1144	795	753	777	1044	1218	1151	1130	740	1387	1317
	A				B				C			

Table 1.

Angle of Incidence to the Collector. With the greenhouse oriented to the south, we can begin examining what happens to solar energy when it reaches the glazing. The sun's rays are most effectively transmitted through a clear material when the angle of their intersection with the surface of the glazing is 90°. This perpendicular is called *normal* (Fig. 3). Because of the earth's rotation and orbit, the sun's rays are normal to any fixed collector surface, like the greenhouse glazing, for one or two moments a year. At all other times of the day and the year the angle of incidence is not normal, or less than optimum. To average the angles of incidence for optimal solar collection using the altitudes of the sun at solar noon, see *The Charts* (Appendix B). You need to know the latitude of the site; the *altitude*, or the height of the sun from the horizon, and the tilt, or angle, of the collector measured from the horizontal. The angle of incidence is the *difference* between the intersection of the solar angle and normal (Fig. 4).

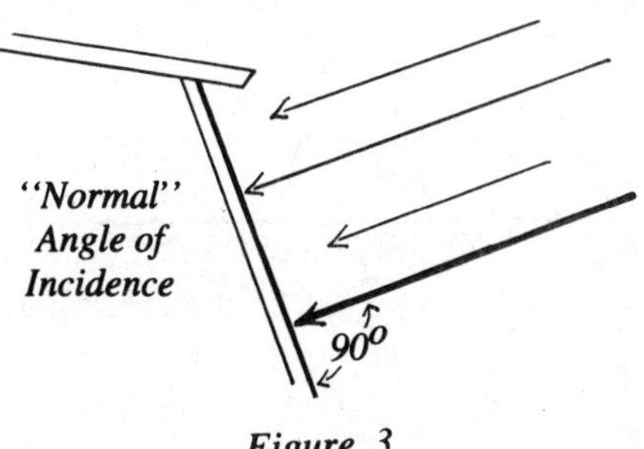

Figure 3

In solar greenhouse design, it is important to get the most energy through the glazing in the winter and to reduce the solar gain in the summer. You can do a great deal to control the heat in the greenhouse by the tilt of the glazing. The chart on p. 14 illustrates energy transmitted through double south-facing glass at various collector tilts. At every latitude, winter collection is optimal and summer is minimal at the steeper tilts (75° and vertical, Fig. 4, p. 14).

SOLAR TRANSMISSIONS THROUGH TILTED DOUBLE SOUTH-FACING GLASS ON A CLEAR DAY

Ground Reflection Assumed at .2
Numbers are BTU/square foot per day

Tilt	30°				45°			
	Solstice		Equinox		Solstice		Equinox	
	Dec.	June	Sept.	March	Dec.	June	Sept.	March
Latitude 36°	1386	1910	1859	1938	1591	1630	1854	1936
Latitude 40°	1216	1971	1811	1890	1426	1719	1838	1922
Latitude 44°	1026	2022	1748	1829	1229	1799	1806	1894
Latitude 48°	835	2062	1671	1753	1026	1868	1759	1849
Tilt	**75°**				**Vertical**			
	Solstice		Equinox		Solstice		Equinox	
	Dec.	June	Sept.	March	Dec.	June	Sept.	March
Latitude 36°	1661	816	1475	1539	1527	446	1102	1146
Latitude 40°	1535	940	1527	1598	1435	527	1191	1243
Latitude 44°	1362	1062	1563	1641	1292	628	1263	1324
Latitude 48°	1173	1178	1581	1666	1130	740	1317	1387

Table 2.

Transmittance through Multiple Glazings. The number of glazings that cover a collector is extremely important. Each time you add a glazing to retain heat, you lose a substantial amount of light. Most of the glazings used in greenhouses transmit 80-90% of the light that strikes their outer surface under ideal conditions.

100% of the light strikes surface.

1) 90% transmission rating through one surface

2) 90% through 2nd layer
<u>X.90</u> transmission rate
81% transmission through two surfaces

3) 80%
<u>X.90</u>
72.9% transmission through three surfaces

4) 72.9%
<u>X.90</u>
65.61% transmission rate through four surfaces

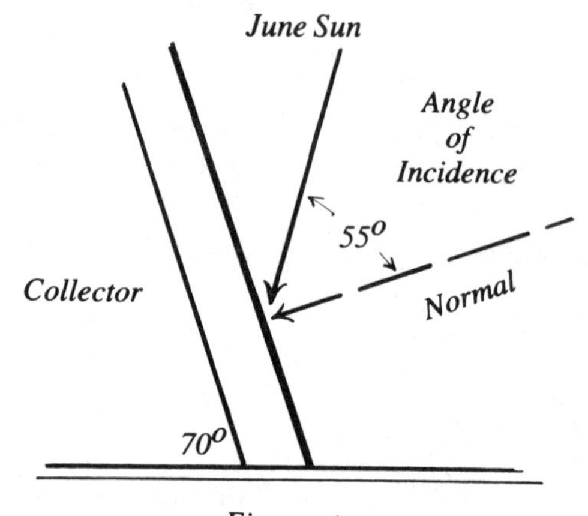

Figure 4

In actuality, even less light than this will get through four surfaces but you can see what's happening; add more layers and you won't have enough light left to grow a mushroom.

Considerable research has been done to determine the proper number of glazings for a solar greenhouse. The general consensus is that *two* is the best in most instances. One layer is appropriate in very warm climates, like the southern United States. Three is often justified in very cold climates where constant snow cover is the rule in winter, because the snow makes up in reflection what the third glazing loses in transmission. However, there are exceptions to this rule. For instance, if movable insulation is planned, fewer layers of glazing may be appropriate.

Reflection of Light. As the north wall of an attached greenhouse is solid and does not transmit light, it is necessary to reflect (or bounce) some light from it in order to duplicate the naturally diffused light that a plant would receive outdoors. If this is not done, the plants can become abnormally *phototropic,* or light-seeking, and will not exhibit healthy growth patterns. In a freestanding greenhouse, the north wall can be tilted to reflect more light to the plants.

Dark and opaque surfaces combined with heat storage are required in the solar greenhouse to absorb and conserve heat, while light and clear surfaces are essential for healthy plant growth. The solution to these conflicting needs is a compromise. Some surfaces will absorb while others will insulate and reflect. The reflective areas can be placed directly behind the plants on the north side of the greenhouse. Table 3 shows the reflective properties of various building materials in the visibile light range.

Reflectance of Commonly Used Building Materials

Material	Reflectance (Percent)
WHITE PLASTER	90 - 92%
MIRRORED GLASS	80 - 90%
MATTE WHITE PAINT	75 - 90%
PORCELAIN ENAMEL	60 - 90%
POLISHED ALUMINUM	60 - 70%
ALUMINUM PAINT	60 - 70%
STAINLESS STEEL	55 - 65%

Table 3

Heat Use

The Greenhouse Effect. When solar radiation in the form of *short waves* passes or is *transmitted* through the clear glazing of the greenhouse the energy hits objects inside. The short waves are changed to a *longer* wavelength (Fig. 5). This longer wavelength does not readily return through the glass; it is blocked. This is called the *greenhouse effect* and the result is *heat*. If this principle isn't clear, think of your car on a clear winter day. If you leave the windows rolled up and go shopping for an hour, when you return the interior air temperature and the seat covers will be warm.

By the way, *temperature* is not *energy*. It is a measure of the *effect* of energy on a substance. It is quite relative to other temperatures. For instance, imagine the temperature of a 50° room in winter. It seems cold, doesn't it? Now picture yourself walking into that room after having been out in a sub-zero blizzard for an hour. The image, and the reality, is warmth. This is an important concept to remember when we discuss how heat supports life in a greenhouse.

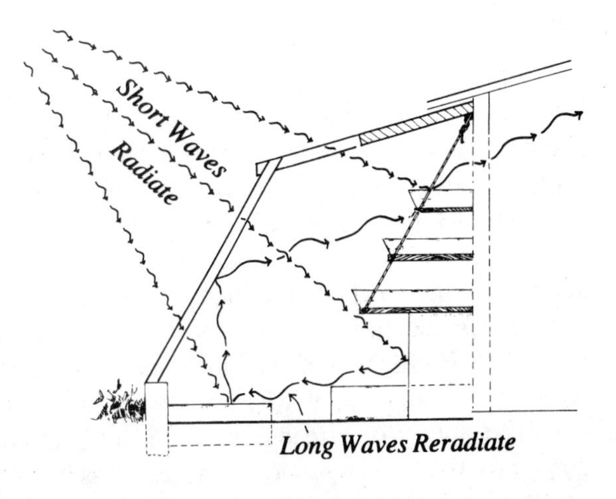

Figure 5

Passive and Active Solar Applications. A *passive* solar energy system is one in which all heat is distributed by natural means. An *active* solar system uses mechanical devices, such as fans or pumps, to distribute heat. Many of the applications you will see in this book are being called *hybrid* systems; that is, they employ some passive techniques and some active mechanisms. For an example, the greenhouse is a *passive* solar collector, and a fan that distributes the heated air to the house is *active*.

The Degree Day. This is the unit of measurement used to describe the heating needs for an area. Degree days are obtained by subtracting the *average* daily temperature from a 65°F base. If the high temperature reached 50° and the low was 20°, then the average temperature would be 35° (50 + 20 ÷ 2 = 35); subtracting the average temperature (35°) from the 65° base makes it a 30° day (65° – 35° = 30°). Degree days are totalled throughout the winter heating season and are used for determining the size of solar heating systems or conventional equipment. They can also be used to present a relative picture of heating needs in different parts of the country. General climatic conditions can be categorized according to totals of winter degree days.

0-2000 Degree Days--------------------------Warm
2000-4000 Degree Days --------------------Moderate
4000-6000 Degree Days-----------------------Cold
6000+ Degree Days--------------------------Very Cold

The degree-day measure is convenient but other seasonal factors must be kept in mind. For instance, both Seattle, Washington, and Albuquerque, New Mexico have about the same degree days, 4,300. Seattle is cloudy and wet in winter, Albuquerque is sunny and dry. The latter obviously will be a better area for solar applications, if all other factors are equal (see Appendix D, p. 181.)

Thermal Characteristics of the Greenhouse

The Second Law of Thermal Dynamics says that, unhindered, heat will always move to a colder area regardless of the direction it has to go. It is indifferent to "up" or "down," "inside" or "out." Greenhouses lose heat in ways that are both positive and negative to overall performance. Here are some of the general principles of thermal dynamics which will help you to minimize negative effects and maximize positive ones—in other words, manage heat loss. These three forms of heat transfer are occurring constantly.

Conduction. Conduction is heat movement through a solid mass, between bodies in contact with one another. This idea becomes clear when you grasp the handle of an iron skillet from a hot burner without a protective pad. The iron in the skillet is an excellent conductor of heat and it doesn't take very long for that heat to move from the bottom of the skillet up the iron to the end of the handle. Ouch! Because the element heating the pan is so much hotter than the pan itself, and the handle of the pan in turn so much hotter than your hand, the conduction of heat takes place rapidly.

All forms of heat loss take place at a *faster rate* when the temperature difference between the two areas is greater. In the greenhouse, heat losses to the outdoors will be less during the daytime than at night because the temperature difference between the interior and exterior surfaces is less during the daytime.

The ability of a material to conduct heat is called its *thermal conductivity*. The overall thermal conductivity of a wall section is expressed as its *U-value*. In the greenhouse, the primary materials that are conducting heat outdoors are the glazed surfaces, the framing members and the poorly insulated walls and foundations. We can slow down (but never stop) the flow of heat by lowering the U-value of a surface. The inverse of the U-value (I/U) is called the *R-value* and gives us a measure of resistance to heat transfer. Increased resistance to heat flow is accomplished by air pockets in an insulating material. The more trapped air pockets, the better the insulator, and the higher its R-value. To reduce conduction losses through the surfaces of the greenhouse, we raise their R-values.

Here are the R-values of some common building materials rated for a one-inch thickness:

Material	R per inch
Expanded Polystyrene	5.26
Loose Cellulose	3.70
Batt Fiberglass	3.17
Dry Earth	2.25
Soft Wood (framing lumber)	1.25
Common Brick	.20
Concrete	.08

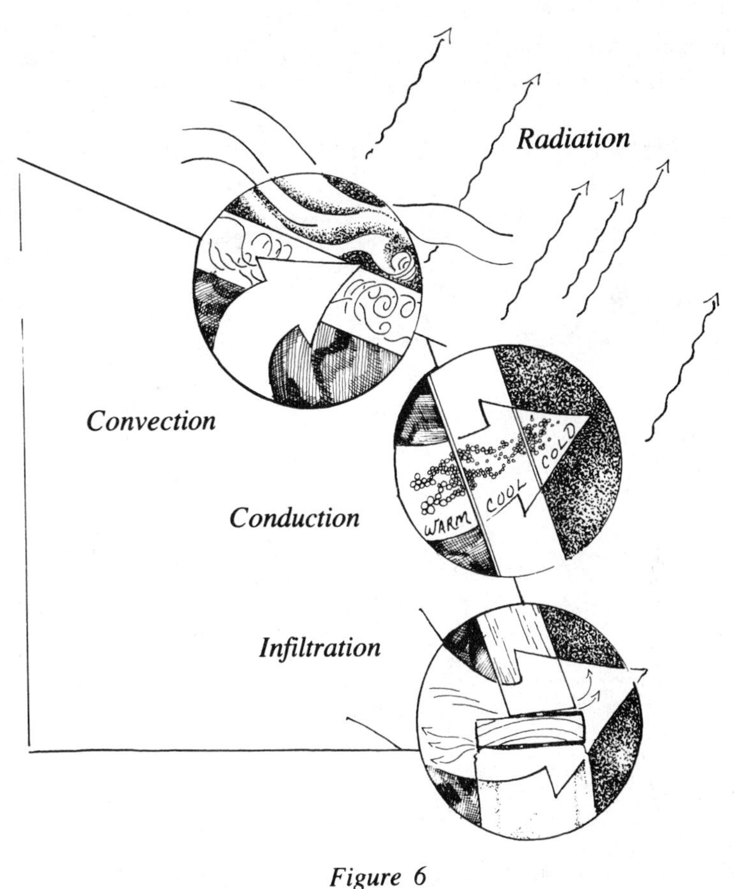

Figure 6

Conduction heat loss through the glazed surfaces of the greenhouse makes all the other surface losses seem almost insignificant. The U-value of a single layer of glass is 1.13, so the R is 1/1.13 or .88. Two layers of glass with 1/4'' air space between panes has a U of .65 or an R of 1.5. So by double-glazing the largest surfaces in the greenhouse, heat losses through the glass are cut almost in half. Adding an insulating barrier with an R-value of only 2 to cover the glass at night produces the following results:

Double glass	R 1.5
Insulating Blanket	R 2.0
Air film between blanket and glass	R 0.6
	R 4.1 TOTAL, or over four times the insulating value of a single layer of glass, less than 1/4 the heat loss of the single-glazed greenhouse.

Convection: Air is a fluid. (Note: as opposed to a solid, fluid is *not* the same as liquid.) Hot air rises and cold air falls; this is another basic principle of thermal dynamics. When a fluid (such as air) is heated, the distance between the molecules increases, and a given volume of the fluid will thus be lighter and will rise. As it cools, the distance between the molecules decreases; the fluid becomes heavier and gravity pulls it down. This action is called *convection*.

Convection patterns in the attached solar greenhouse can be beneficial or harmful. During nighttime periods, for instance, when the clear glazings are colder than the thermal mass of the greenhouse, a *convection loop* is established that contributes to heat loss (Fig. 7). Warm air rising from the thermal mass is pulled across the glazed surfaces by the heavier cold air. So the warmer air loses its heat more rapidly, through conduction, to the outside. The greater the temperature differences between cool glazings and the warm objects, the faster the convection currents move and the more heat is lost. Insulating barriers such as the transversely mounted blanket shown in Fig. 44, p. 49, help to break up a large convection cell into two different temperature zones and in doing so greatly reduce convection heat losses.

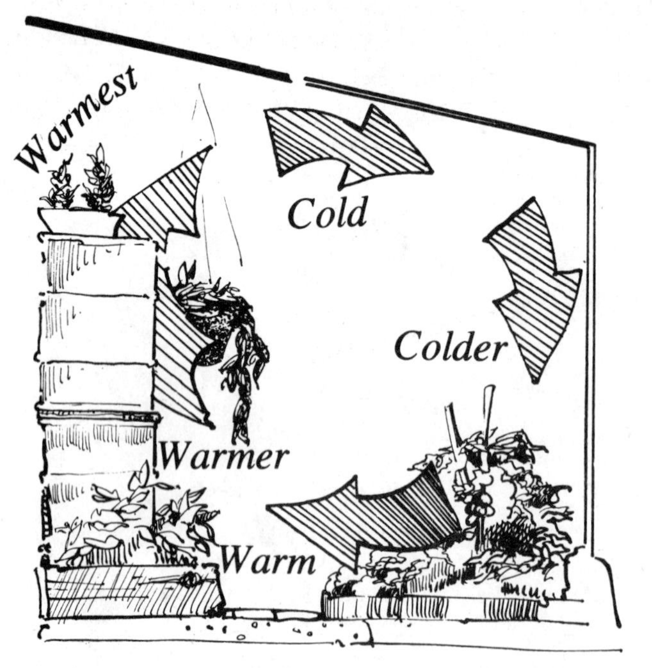

Nighttime Air Movement in Greenhouse.

Figure 7

Both convection and conduction heat losses are higher when outdoor convection patterns, *wind currents,* are increased. Wind blowing across the surface of the greenhouse will cause the outer surfaces to be cooled and more heat to be conducted through them. Wind also creates a high-pressure area on the outside and low pressure inside. The result is faster air leaks by *infiltration* through any cracks or loose joints you might have.

Convection of warm air from the greenhouse to the adjoining home, on the other hand, is a major benefit of the attached unit; it is partially through this daytime process that the solar greenhouse becomes a winter heating system. The sun is the power source and the home is the lucky recipient in this partnership. The convective loop is established on clear or partly cloudy winter days when the greenhouse air temperature rises above that in the interior of the home. Through high and low openings (vents, doors, and windows) to the home, a natural convection cycle is created that will run as long as there is sufficient solar radiation into the greenhouse (Fig. 8).

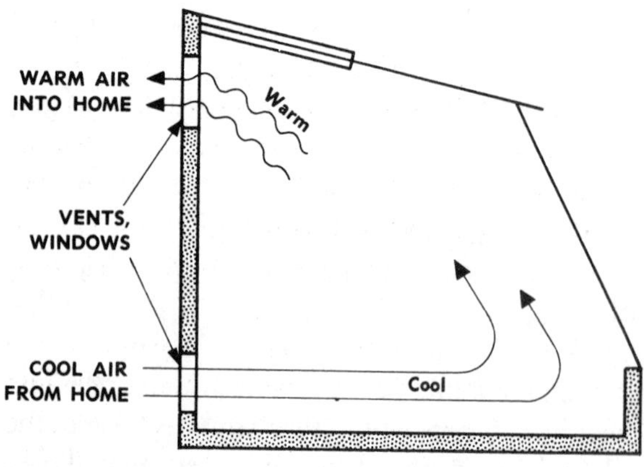

Daytime Air Movement in Greenhouse.

Figure 8

Radiation. Radiation is energy transmitted as electromagnetic waves directly from one body to another. The energy transfer takes place without a medium until the waves are reflected by a radiant barrier or absorbed by a solid. Imagine yourself standing a distance from a bonfire on a cold winter night. You can probably remember your fire–facing side being quite warm and your other side freezing. That's radiant transfer. If you want to get warmer, you either have to get closer to the fire to capture more radiation (and convective heat), or rotate yourself slowly.

Radiation is a two–way street. While all bodies transfer radiant energy back and forth, the net difference is from the warmer to the cooler body. So, in the example just given, if there is a car close to you and its metallic surface is colder than your backside, it is going to receive radiant energy from you, even as it is giving off radiant energy to the clear sky above, which is colder than any of the other objects.

On a winter night, radiant heat loss from the greenhouse through the clear glazings to the *night sky* is substantial, as much as 40% of total heat loss on very clear nights. Glass and plastic glazings absorb the radiation and their temperature becomes higher as a result. This increases conduction losses through the glazing. A simple foil barrier is effective in reducing the heat transfer out. In the design section, some methods of slowing radiation losses are given.

Radiation *heating* is a primary principle in most passive structures, and solar greenhouses are no exception. For the greenhouse, heat absorbed by thermal mass within the structure (water drums, masonry walls, soil) is radiated directly to the plants at night when their surface temperatures drop below the temperature of the thermal mass (Fig. 9). If we surround the plants with warm radiant surfaces, they can tolerate much lower air temperatures.

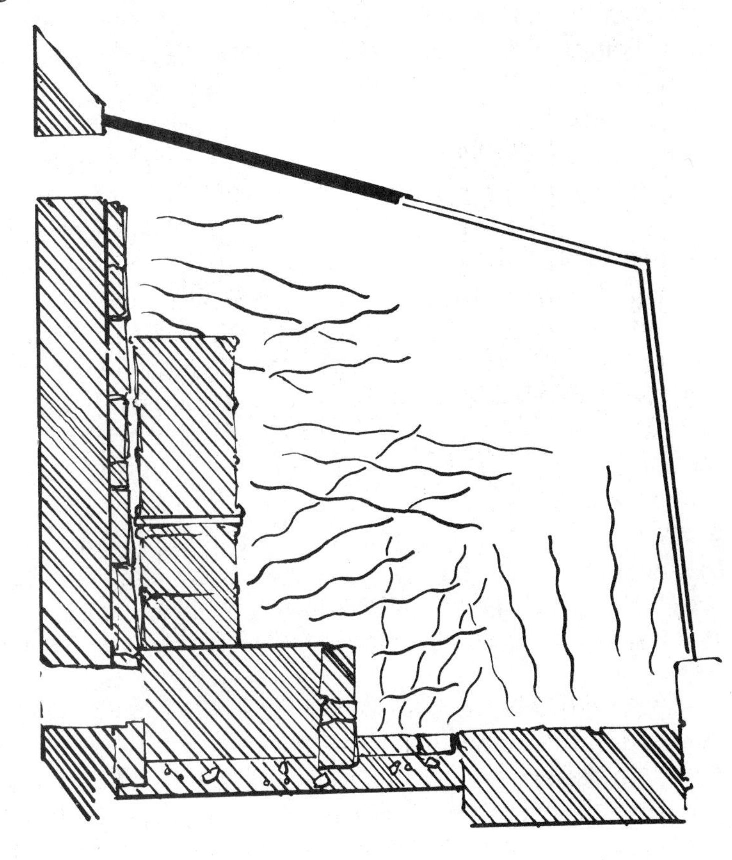

Figure 9

In a greenhouse-home combination (see Appendix E, p. 183) heat from the greenhouse can be conducted through the adjoining wall to the home interior. The entire wall becomes a low-temperature radiant heater. This is the best of all heating systems because there are no hot spots, no noisy fans, and absolutely nothing can break down.

To effectively use radiant heating for the greenhouse and the home, mass must be properly sized, placed, and colored. Dark-colored objects and materials in the greenhouse *absorb* the energy from the sun during the day, and their temperature is raised. If these objects have sufficient *thermal mass*—the capacity to absorb, store, and distribute appreciable amounts of heat—then they will, in effect, capture some free energy for later use. Think about a hot-water bottle. You fill the rubber bag with very hot water and place it between the cold sheets of your bed. When you go to bed an hour later your toes will have a nice warm place to snuggle. The device conducts its heat to you and the bed for many hours. In the morning when you wake up, it may still be lukewarm. Heating by radiation (and conduction) with thermal mass in the greenhouse works on the same principles. A mass of material in the greenhouse soaks up heat from the sun during the day, then slowly releases that stored up energy back to the greenhouse or to the adjoining structure at night. Technical tomes have been written on how to properly size thermal mass to the heat loads of any particular building in any particular climate (see *Bibliography*). The factors used include conductivity, surface to volume ratio, color, mixing ability, and placement. In the design chapter we give you some "rules of thumb" that apply to attached greenhouses and serve as a point of departure for the design of new greenhouse-home combinations.

Condensation and Evaporation. The moisture content of an air-water vapor mixture, when expressed as a percentage, is called relative humidity. Warmer air can hold more humidity than cooler air, which is why the indoor relative humidity is usually higher in summer months than in winter. Although humidity can be a real benefit in the greenhouse, it can create some problems because water condensation on different surfaces contributes to their deterioration. For example, when convection loops (Fig. 7, p. 18) pull the humid greenhouse air across the colder, clear-glazed surfaces, condensation can occur on the inside glazings.

In winter, humid air carried by convection from the greenhouse to the house will be appreciated if the unit is attached to a well-insulated room. If the adjoining room is poorly insulated, the warm greenhouse air will form condensation on the north wall of the home. Ideally, humidity in the greenhouse should range from 30% to 70%. Not enough moisture in the air dries out plant tissues; too much moisture promotes disease, especially in combination with high temperatures. A plant must be able to lose moisture by *transpiration* (the release of water vapor) to keep from overheating. Air movement through ventilation is the most effective way to control excess humidity. In the winter, venting the humidity into the house is helpful, as most homes are too dry. To control summer humidity and excess heat you must have an efficient system for *moving air* from the greenhouse to the outdoors. (See Appendix H, *Vent Sizing*, p. 189.)

When water evaporates, one pound (a pint) will absorb about 1000 BTU's of heat energy. So evaporation is a very effective and widely used form of cooling in parts of the world that experience hot, dry weather. With natural ventilation (no fans) the greenhouse will often act as a natural evaporative cooler for itself. Through transpiration the plants in the greenhouse act as so many swamp pads: air flowing past them absorbs the moisture and is cooled. This works quite well in the western United States through the summer and in most parts of the country during spring and fall. In humid locations, evaporative cooling of greenhouses can still be effective and is the conventional method for cooling commercial greenhouse structures. Unfortunately, the higher the relative humidity, the more difficult it is to get evaporation to occur. The small home greenhouse in a warm, humid climate can use a small fan to aid in summer cooling.

The Site

The meaning of the term *trade off* will become apparent when you begin to select a site for your greenhouse. All the conditions are not likely to be ideal, but it is important that positive factors are emphasized and detrimental ones kept to a minimum.

Your first step in choosing a site is to determine how your home and property are aligned in relation to solar movement and other natural elements. Stand on the south side of your house. Where did the sun come up today? Where will it set? What will its rising and setting positions be on December 22nd and June 22nd (the solstices) relative to your south wall? These are the basics of solar design. This is where it all begins. All of the natural considerations apply to independent as well as to attached solar greenhouses. Use *The Charts*, Appendix B, of this book as aids in visualizing orientation, sun movement, and obstructions at your location.

Sun Movement and Building Orientation. The sun is constantly changing its path through the sky, dropping low on the horizon in the winter and rising to an overhead position in the summer (Fig. 10). A solar greenhouse differs from most solar applications in that it is not necessary to obtain the maximum intensity and duration of sunlight throughout the year. It should be designed and located so that it receives the greatest possible amount and intensity of direct sunlight during the winter, when daylight hours are few, and less light in the summer when overheating is a problem. The *photoperiod* becomes particularly important for plant growth in the greenhouse during the winter. Because the days are so short from October to March, both the plants and the heat storage features of the greenhouse need every available minute of sunlight. Designing in accordance with sun movement patterns gives the solar greenhouse automatic advantages over conventional greenhouses for winter heating and summer cooling.

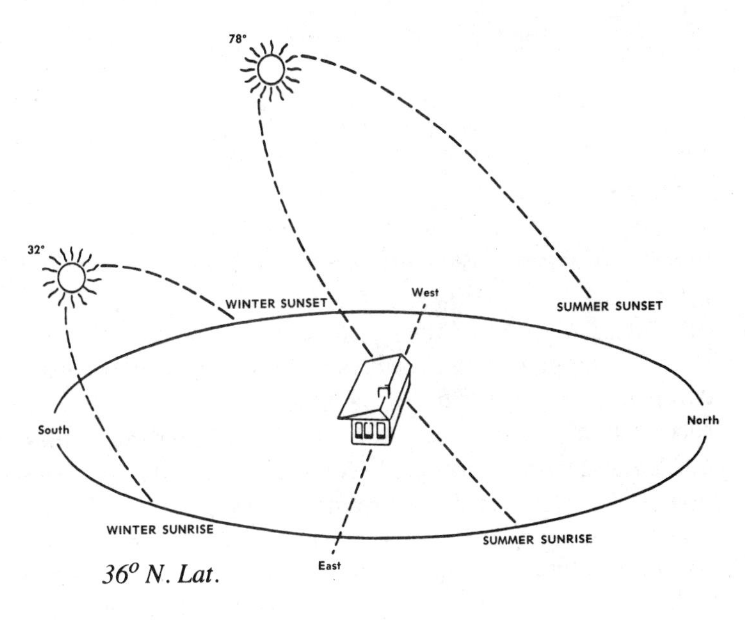

Figure 10

Estimating South

There are many ways to estimate south. One is to find Polaris, the North Star, and place two stakes in the ground about three feet apart that align perfectly with the star. That line will be true north-south axis. If you can't find Polaris, consult a Scout or look at a star chart.

Another old Scout trick uses a conventional wrist watch (not a digital readout). On a clear day, around 8 A.M. or 4 P.M., point the hour hand of the watch directly at the sun. Keep the watch level. Halfway between the hour hand and the 12 o'clock position on the watch will approximate true south. Be sure you're on *sun time*—not daylight savings time.

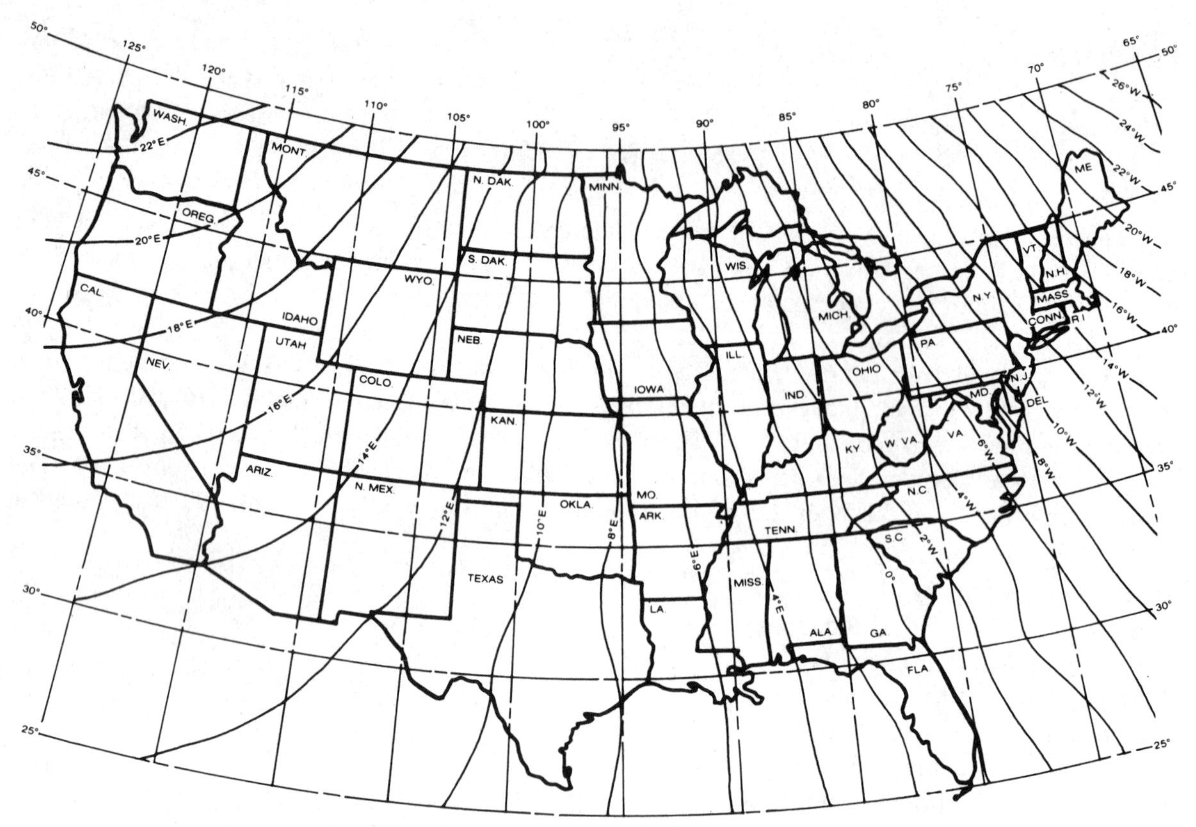

Isogonic Chart shows magnetic deviations for continential U.S.

Figure 11

A solar greenhouse requires at least a *partially southern exposure*. Find magnetic south by using a compass. A survey map or the chart above (Fig. 11) can tell you how many degrees east or west (declination) from your site true south is. Add or subtract these degrees to find true south. For instance, if the deviation in your area is 12°W (west), true south is 12° west of the south pointer on the compass. When you establish true south, determine how far from a perpendicular to south your house wall is. This east-west axis will be the north wall of the greenhouse and it can be as much as 45° off true east-west without losing an appreciable amount of winter sunlight, although no more than 15° is optimum. The diagram below shows some good orientations for attached solar greenhouses (Fig. 12). If the corner of your home points south, consider a corner location (Fig. 13).

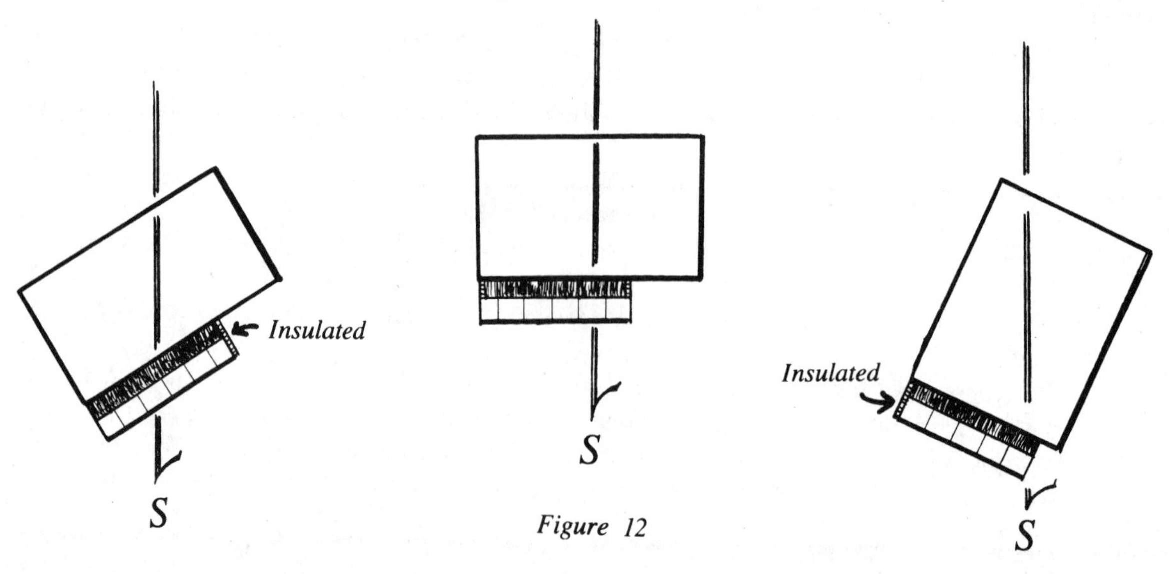

Figure 12

Obstructions. Anything that blocks light from the greenhouse may be considered an obstruction. In examining an obstruction, its orientation and size are the two most important factors. Where is it in relation to the south face of the greenhouse? How much of the time is it blocking the sun and during what season (Fig. 14)? A deciduous (leaf-shedding) tree that partially shades the greenhouse from the late afternoon summer sun could be an asset. A twenty-foot evergreen ten feet south of the unit is a serious problem. *The Charts* in Appendix B can help determine how much sun will be blocked at various times of the year. For maximum solar gain the greenhouse needs to be *unobstructed* from 9 A.M. to 3 P.M. during the winter photoperiod. Shading in the early morning or late afternoon isn't as costly to thermal performance as it is to the plants; they need all the light they can get during the short winter daylight hours.

For best midwinter operation, no more than an hour or so of midday sun can be lost to obstructions. In urban areas the possibility of a neighbor planting a tree or adding another story to his home directly in front of your greenhouse is something you should consider. It may surprise you, but in most of the United States you don't have any rights to sunlight. Right to the sunlight that falls on your land is not considered part or parcel of rights that usually accompany ownership of the property such as easement or access. Sunlight is generally considered *incorporeal,* or without body, so has no

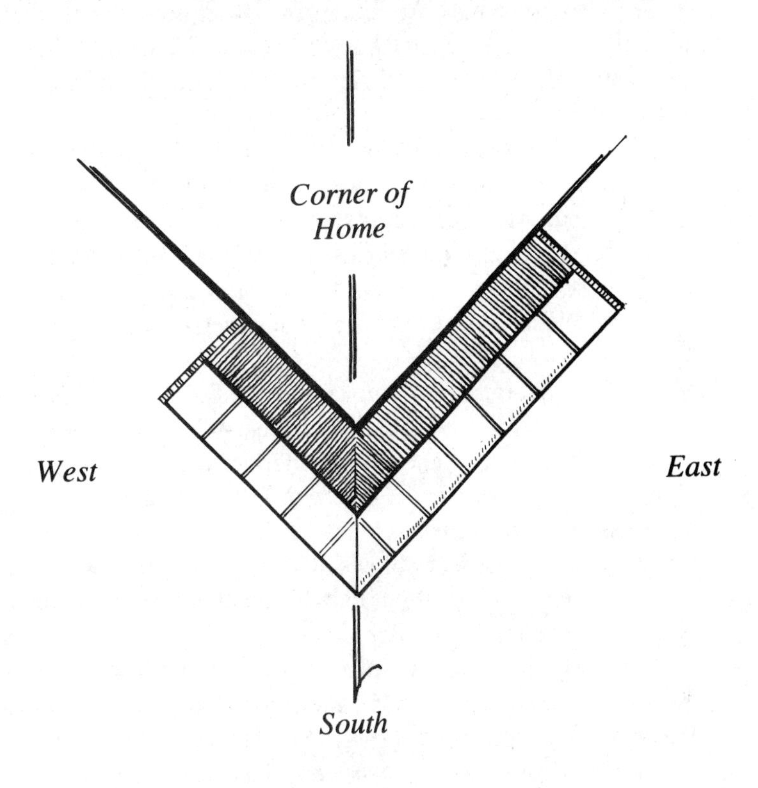

Figure 13

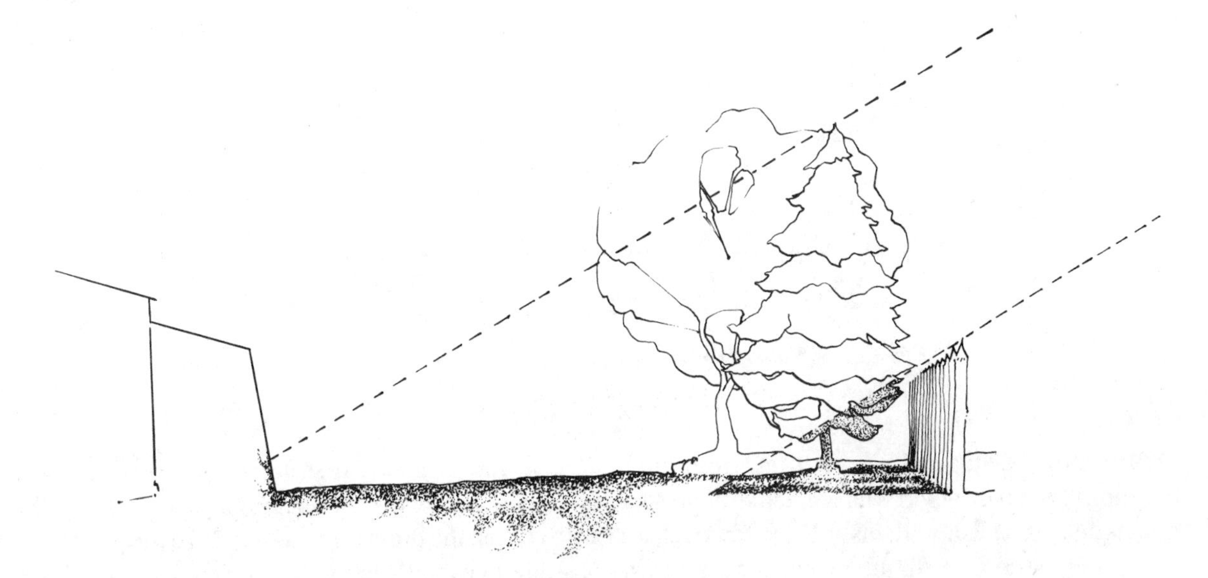

Figure 14

legal definition under law. Therefore, legally it doesn't exist. If this bothers you, work with your local solar energy association to change things.

Your home may present a sun obstruction on the east or west side (see Fig. 12, p. 22). While it might block some of the light, particularly in the summer months, it will also be blocking the wind and acting as an insulating barrier against the elements.

The important thing to remember is that an obstruction can be a positive or negative factor. You can compensate for some obstructions by proper bed layout. In other cases, exterior reflectors can make up in intensity for light blocked by an obstruction. Remember that whenever light is sacrificed, the performance of the greenhouse is altered. Try to achieve the full winter photoperiod and at least ten hours of summer sun.

Local atmospheric conditions affect the light the greenhouse receives at various times of the year. In many mountainous areas a winter morning is more likely to be clear than a winter afternoon. It is advisable in those regions to have a greater amount of clear wall facing east than facing west. Generally, some eastern clear glazing makes sense in any locale and is usually preferable to a western one. The greenhouse needs these early morning rays after a cold winter night. Joan Loitz (Chapter VII) claims that morning "is when the plants do their growin'. Give them eastern light."

Wind. The natural flow of prevailing winds can be used to your advantage in the design of the greenhouse. In many parts of the country the summer-winter wind patterns will vary as much as ninety degrees. The southwestern United States has a pattern of winter winds from the northwest and summer breezes from the southwest. By mounting a low vent in the southwest corner of the greenhouse and a high vent in the northeast, the prevailing summer winds are used in natural cooling. There is no national map that can tell you which way the wind blows at various times of the year in your neighborhood. Local topographical features, trees, and buildings can drastically affect wind patterns. The best guide is local experience with typical conditions ("You don't need a weatherman to know which way the wind blows."—Bob Dylan).

In designing your unit, place the lower vents for summer cooling on the side of the greenhouse *facing* the summer prevailing winds. Also, try to locate the *exterior* greenhouse doorway on the *opposite side* of the prevailing winter winds (Fig. 15). In this way the entrance is partially protected from drastic heat losses when opened during the cold period. This is of special importance in a greenhouse that has no doorway to the home.

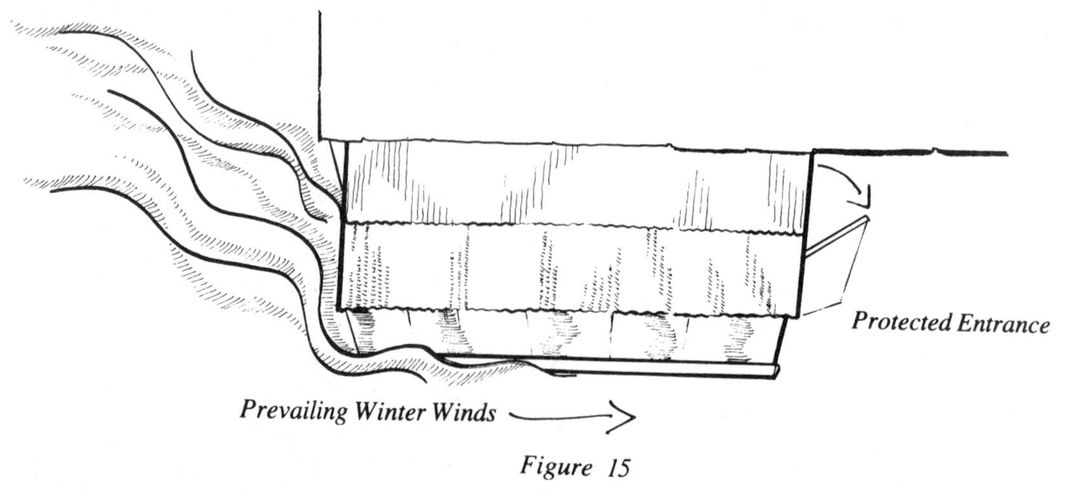

Figure 15

Drainage. Adequate drainage away from the home and the greenhouse structure is essential. Most buildings are constructed with a gradual slope away from them for runoff. When the extra roof area of the greenhouse is added, will this still be effective? Be sure to check the drains and gutters from the house and where they terminate. In urban locations it may be possible to connect to existing drainage lines. Try to have the ground runoff from the greenhouse follow the existing pattern of drainage. Some pick and shovel work may be necessary to facilitate this.

Percolation, the rate at which water can flow downward through the soil, is particularly important if you plan to have ground beds in the greenhouse. Although correct watering procedures would never allow saturation of the beds, accidents (such as leaving the hose on overnight) do happen. When soil conditions at the greenhouse site are not conducive to good percolation, you can add several inches of coarse sand or gravel to the ground level of the unit to aid drainage. Water accumulating under the floor is not beneficial to either the plants or the thermal dynamics of the greenhouse. It lessens the effectiveness of any insulating barrier.

Utilities. It is convenient to have water and electricity available at the greenhouse site. A water faucet cuts down on the manual labor involved in hauling water to the plants, but because plants in a greenhouse do not require as much water as they would outdoors, their needs can be accommodated by hand. If a faucet is located at the site, plan to build it into the unit. In the spring and summer you can extend a hose through the door or vents for watering the outdoor plants. It is also possible to get an adaptor for indoor water outlets and run a hose to the greenhouse through a door or window.

An electrical outlet is also convenient but not essential. It is enjoyable to have light for nighttime work, but difficult to justify the expense of the electrical power needed to light the structure in terms of the additional food it could produce. Also be sure to check the potential site for underground utility connections *before* you dig anywhere. Driving your spade through a 220-volt service wire can be a shocking experience.

The main point is this: leave or design provisions for utilities if it is convenient and not costly to do, but don't feel that you have to have them in order to have a successful solar greenhouse.

Building Codes. Building codes, inspectors, and permits are strange inventions. Originally intended to be constructive, helpful devices, they can be restrictive, rigid, and generally oppressive to innovative design work. The latest information in the code books about greenhouses was probably written around 1940. In some regions, greenhouses may be considered "temporary" structures (like gospel show tents) and have virtually no restrictions on their construction. In other areas, they may be subject to strict (and obsolete) codes.

The biggest problem for attached greenhouses arises when they come in conflict with the "light and ventilation" section of the Uniform Building Code adopted by many states. The problem is an outdated law that considers the attached greenhouse an agricultural building, not a habitable space. You will have to work with the inspector on a personal basis to convince him/her that the unit is primarily a living addition to the home and should be judged as such. Often difficulties arise over the accepted name the solar greenhouse carries in your area. You may be better off building a "solarium," "atrium," "sun room," or just "enclosing a porch." By finding the right name before applying for the permit you avoid problems and you may also qualify for *tax credits* or *rebates* that apply to solar structures in many states through financial solar initiatives.

The best advice is to find a friend involved in construction and check up on the "mood" of the codes and inspectors in your area. Quite possibly the local inspector will be a considerable aid in your project, giving valuable advice on the strength of lumber, foundation footings, and so forth. If you are in doubt about the local situation, follow the prescribed code to the letter rather than running the risk of violating it. In the long run this will be cheaper than tearing the structure down and starting over.

Greenhouse Configurations

When you start planning the solar greenhouse addition to your home you should consider: (1) the overall efficiency of the design in terms of the amount of heat it gives to the home and food it puts on the table; (2) the architectural integration of the greenhouse with the house—its size and shape, as well as the textures and quality of the building materials used; (3) time and cost—what you can afford to build and how much time you have to maintain it. The best way to make the right decision is to have a broad understanding of the options available.

The model we use for an attached greenhouse is a lean-to or shed roof configuration. It is practical, economical, and easy to build. In considering its exterior shape you are dealing with three primary factors:

- The orientation of your home to true south (see pp. 21 and 22)
- Obstructions (p. 23)
- The optimum thermal relationship between the greenhouse and your home—that is, how certain greenhouse proportions and dimensions affect the collection of light, storage of heat, and thereby the growing conditions and thermal performance of your unit.

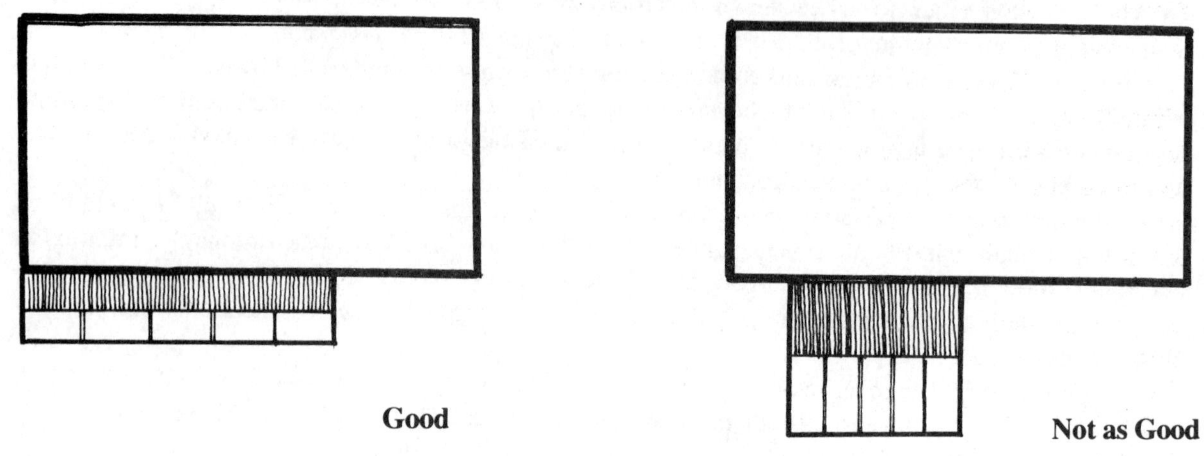

Figure 16

If your home is well oriented and the site presents no serious obstructions, the best shape for an attached greenhouse is a long, rectangular shape rather than a square or "boxy" design. A rule of thumb that has proven successful in attached solar greenhouses is to allow *the length to be about 1½-2 times the width*. A size that has been fairly standard in our demonstration units is 16 feet long by 10 feet wide. This size leaves plenty of growing space with some room for working and relaxation areas. If the width of the greenhouse becomes much greater than 10 feet and the pitch of the roof is shallow, rafters heavier than 2 x 4's must be used and the expense of building increases.

A long, narrow design also allows the home wall to receive light that would not reach it in a boxy greenhouse (Fig. 16).. The more area of shared wall between your home and greenhouse, the better. The mass of the house stores heat, some of which can be used in the unit, and in turn the greenhouse reduces heat loss from the home by acting as a buffer against colder outdoor temperatures. In addition, the greenhouse can more easily supply the home with supplemental heat as it may open onto several windows and/or doors for natural or forced convection. The total amount of heat available to the home from the greenhouse is *directly proportional* to the total amount of south glazing in the greenhouse. By increasing the length of the greenhouse you are producing more heat for your home in the simplest, most direct manner.

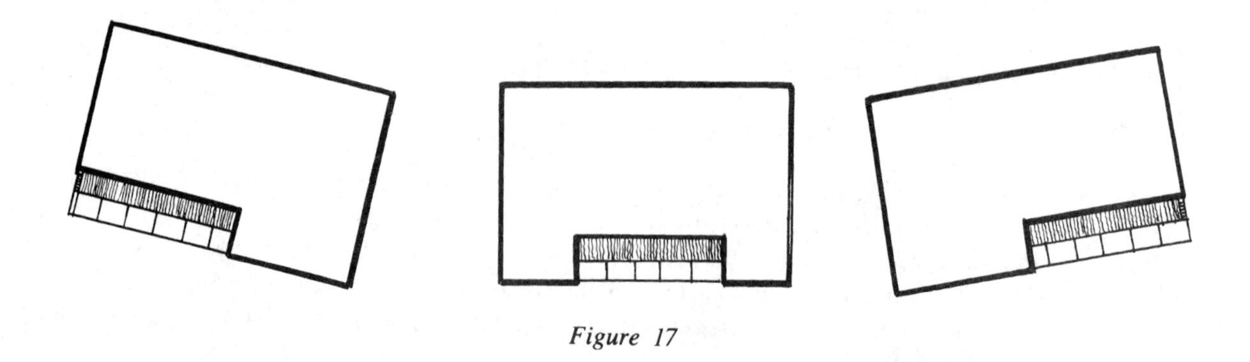

Figure 17

Achieving a symbiotic thermal relationship between the house and greenhouse is as important as exact orientation. Figure 17 illustrates combinations that provide a maximum interchange between the two. If you are lucky enough to have such a desirable home plan for your greenhouse addition, take advantage of it. In Figure 17, the center drawing shows the best possible location for a solar greenhouse in terms of thermal performance. With this site, the house surrounds and protects the unit on all sides but south. Heat losses from the greenhouse are minimized because both end walls are solid and buffered by the house. Solar heating from the greenhouse is optimal because of the large percentage of home wall covered. The indented corner greenhouses will also achieve better thermal performance than the more exposed add-on units shown in Fig. 19. The eastern corner addition is preferable to the western one in most locations, but both are good.

The only drawbacks to the three indented designs can be summer overheating and poor air circulation within the unit. Because the greenhouse is sheltered by the house, additional attention must be paid to ventilation and cross-circulation. Vent-sizing and air movement should be *double* the amount recommended in Appendix H, p. 189. Solar heating will also be improved when the living space of the home has good convective *communication* with the greenhouse (Fig. 18). Large openings between rooms promote air circulation. Closed doors and solid walls prohibit the passage of warm air between areas.

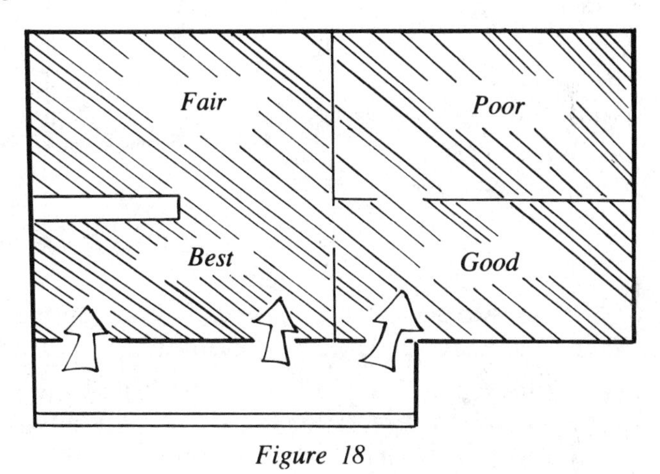

Figure 18

Mobile home owners should follow the same principles given above. Below are recommended layouts based on orientation to true south. The site with the north-south axis on the left is the *least* preferred of the four, as there is less collection area to wall surface covered.

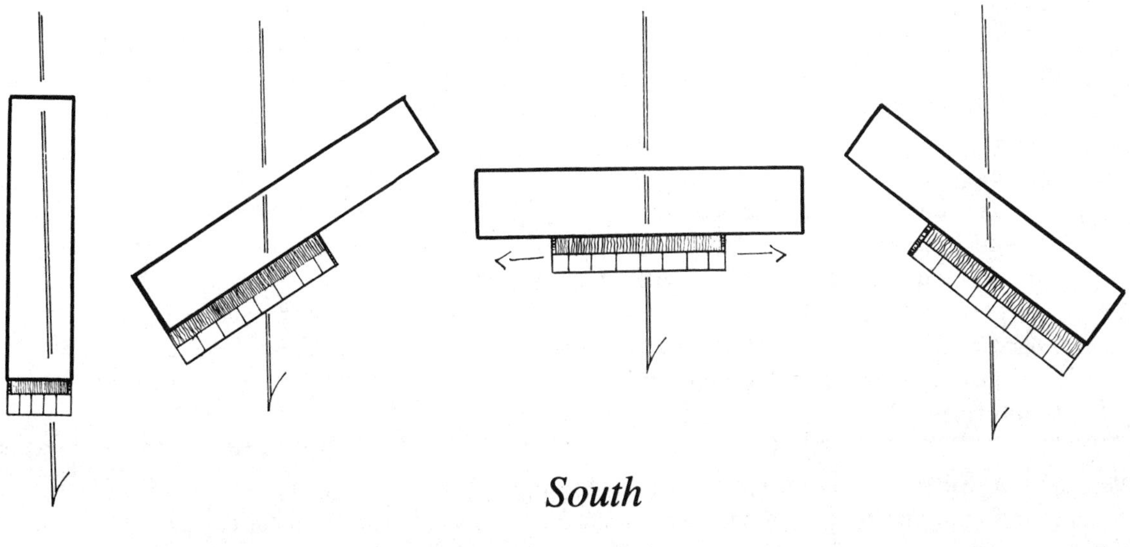

Figure 19

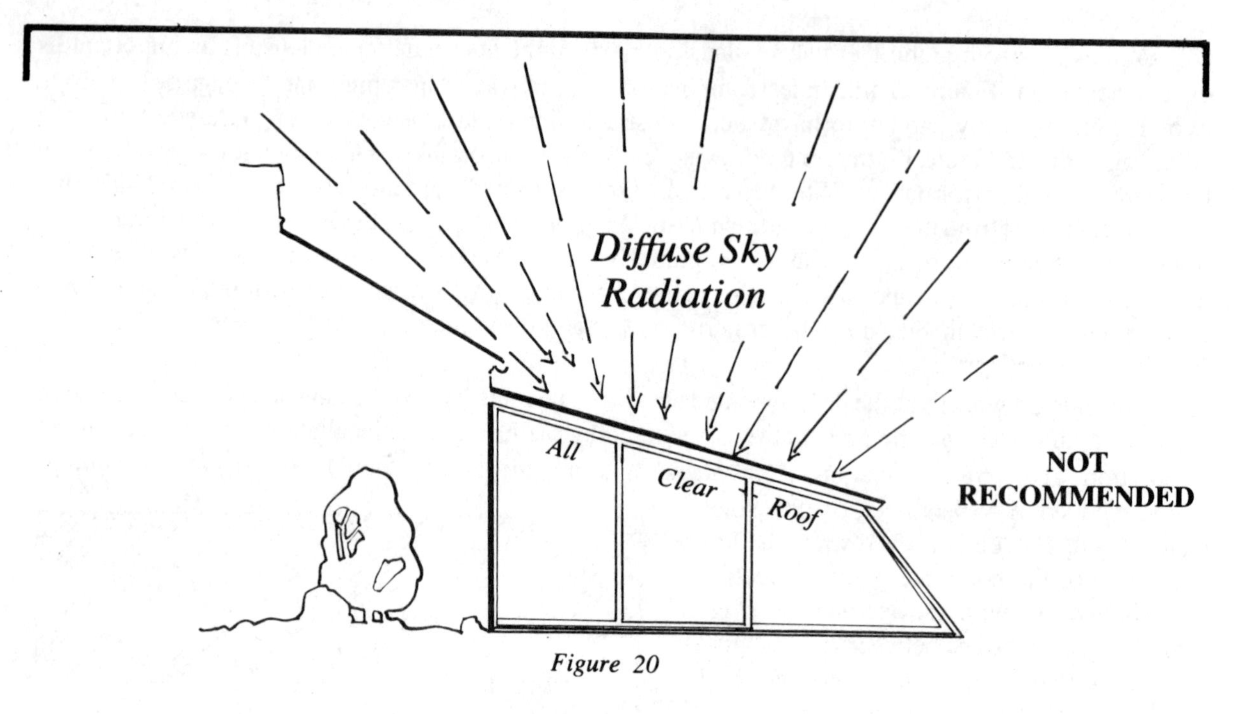

Figure 20

Diffuse Sky Radiation

Although there are numerous disadvantages to the low angle collector, some solar greenhouse publications recommend a clear roof as a way of maximizing collection of diffuse sky radiation (Fig. 20).They report a 50% increase in light in the greenhouse during heavily overcast conditions. Note, however, that increase in diffuse light does not substantially contribute to heating the greenhouse or the adjoining structure, which is a basic premise of the designs in this book. Instead the uninsulated clear area at the greenhouse apex increases heat loss in winter and leads to overheating in the summer. In other words, you're faced with many of the problems of conventional greenhouses. (It's not surprising that conventional greenhouse designs originated in the coastal, cool climates of northern Europe and England. In these regions *unusually* diffuse sky conditions—which also prevail in parts of the Pacific Northwest—make the low-pitched clear roof a sensible design. Here the greenhouse needs the heat in summer almost as much as in winter, and the winters are mild in comparison to harsh inland climates.) The designs we recommend, which combine a near-vertical front face for better winter collection with a partially shaded/insulated roof plane, provide adequate light for plant growth *and* greater thermal efficiency.

The low angle collector is more vulnerable to Mother Nature's little surprises, such as ¾-inch hailstones and falling oak branches. Heavier and safer glazing and framing materials must be used to protect a tilted face aginst these possibilities.

Despite some inherent problems, the 45° geometry has been widely used; in Chapter VIII are several successful examples. Just keep in mind that there are many aspects of solar greenhouse design, and maximizing the collection of diffuse sky radiation is only one of them.

Angle of the South Face. In the solar greenhouse, you are working for maximum efficiency of collection during the winter period and reduced summer transmittance. To achieve that means that the angle of the south face should be nearly perpendicular, called *normal,* to the average "solar noon" angle of the sun during the coldest months. In the northern hemisphere this period is mid-November to mid-February.

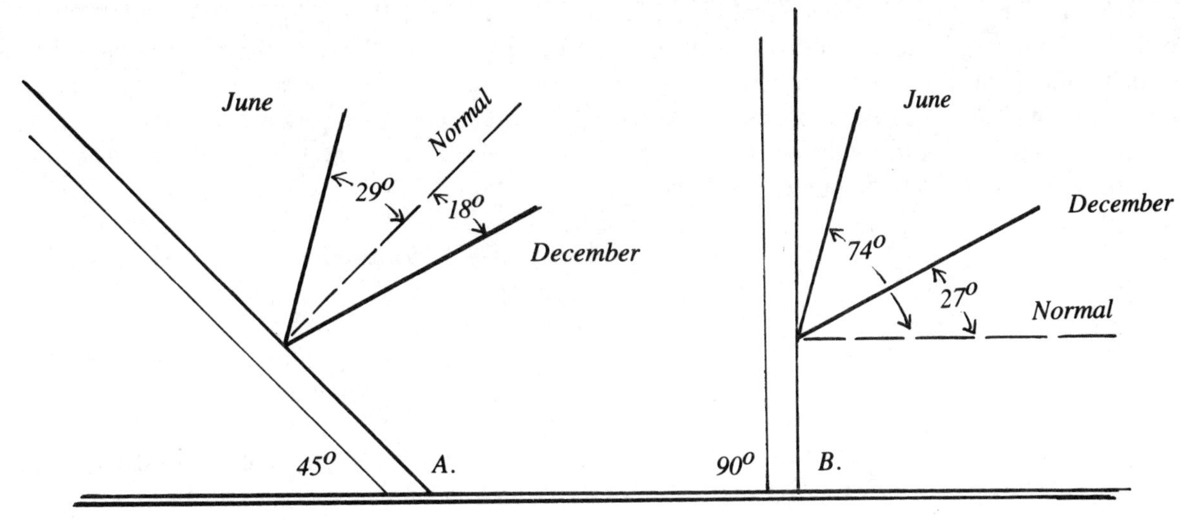

Figure 21

Averaging the winter solar noon angles in the contiguous U.S. shows that the *optimum tilt* for winter collection begins at about 50° in the southern United States and rises to 70° around the Canadian border. Remember, this is at solar noon; at other times of the day the sun is at a lower angle in the sky (see Fig. 10, p. 21). A formula we've used for establishing the tilt of the south face is latitude plus 35°. You can see that if you live north of 45° latitude, the angle approaches vertical. To have exactly the correct angle is *not critical*. The tilt of the glazing can be as much as 50° off normal and still not lose an appreciable percentage of light transmittance. There are charts and technical references that can give the exact percentage of transmission for various collector tilts and sun angles (see Table 2, p. 14, and Appendix B), but the important point is that any clear glazing will transmit the majority of the light that strikes its surface *until* the angle of incidence is greater than about 50°.

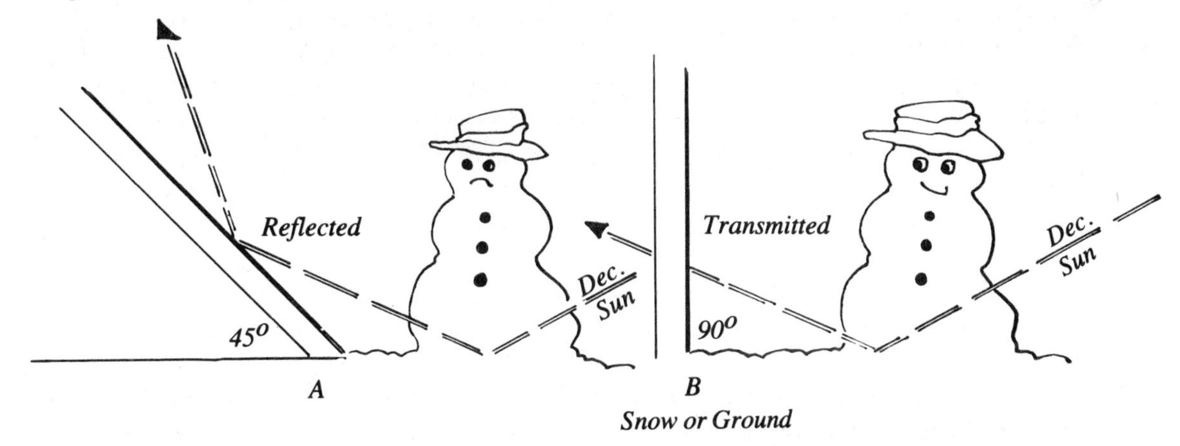

Figure 22

We can compare a collect tilt of 45° with a vertical one (90°) to illustrate the difference in transmittance between a sloped and vertical front face (Fig. 21). Both tilt angles will offer good *winter* collection surfaces, but there the similarity ends. The 45° face will transmit a great deal of solar radiation throughout the entire year. It will collect energy very efficiently right through the summer, just when you don't need the heat in your greenhouse. In contrast, the vertical surface collects well in winter, yet has an angle of incidence of 74° off normal in the summer months. The result is a cooler summer greenhouse because the glazing is *reflecting* the majority of light from its surface in the warm period.

Another advantage to the vertical or near vertical collecting surface is that it transmits radiation reflected from the ground immediately south of it (Fig. 22). The increased collection caused by snow cover or light-colored ground is very important. It can produce up to a 40% gain in solar energy for the collector on a clear day. This is sizable, particularly in severe northern climates that experience constant snow cover for three to four months in winter and need all the extra solar gain they can get. A vertical collecting surface becomes more important the farther north you go.

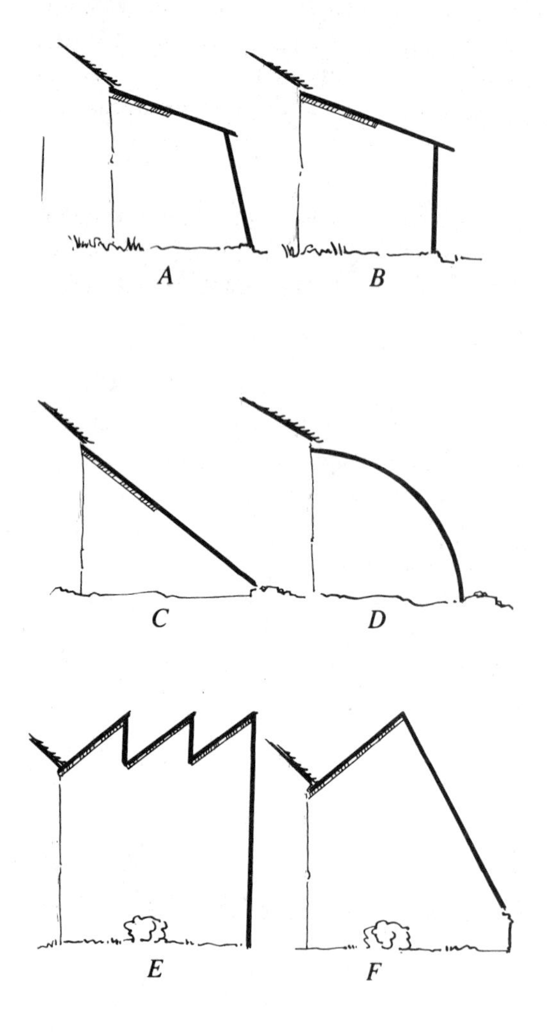

Figure 23

Greenhouse Geometries. Once the orientation, length, and width of your greenhouse have been considered, you can begin to plan the geometry of the unit. We recommend variations on either of the two shedroof geometries pictured (Fig. 23, A and B) and we'll explain in detail the advantages of each in this section. Example A has a front collecting face tilted to coincide with winter sun angles. It is the model we use in the *Construction* chapter. The vertical face of Example B is appropriate for northern latitudes and offers further practical advantages that will be discussed shortly. First, to give you an idea of other options available to you, let's take a quick look at some commonly used designs.

Consisting of a single front plane, design C is perhaps the simplest to build. It should be taken seriously as a very inexpensive *temporary* solar greenhouse. This unit could be applied inexpensively to millions of homes, turning their south walls into solar collectors. Many people find the 45° tilt aesthetically pleasing, and it is extremely effective as an addition to a two-story home. Its most serious drawback as a permanent structure is the lack of usable space in front. This can be remedied by building a low perimeter wall around the base as described in the *Construction* chapter. Another option is to excavate below grade, perhaps to the depth of the home footing, and construct a masonry retaining wall up to or above the ground level.

The "quarter-round" D profile is often seen in prefabricated greenhouse kits using curved metal for the framing supports. The design can be made more thermally efficient using wood framing, but that demands a knowledge of laminating or stressing lumber. In this model the majority of south glazing is within 50° of normal to solar noon altitudes year-round. If you choose the C or D designs and live in an area with warm humid summers, plan on increasing exhaust ventilation, shading, thermal isolation from the home, or a combination of all three, to prevent overheating.

An advantage to the C and D geometries is a large percentage of solar collection area in relation to the floor and home-wall area covered. The corresponding disadvantage is devising a practial, effective method of decreasing heat loss through all this glazing at night (see the section on movable insulation, Chapter V, p. 44).

The two geometries pictured in E and F have been used to increase winter collection in severe climates and for attaching units on homes that have a less-than-ideal orientation. For instance, if the south wall of your home is on the property line or is heavily shaded, E or F can be added to the east or west sides of the house. As you can see, both designs feature huge expanses of steeply tilted or vertical glazing in relation to the floor area covered. They are excellent solar collectors in winter. Note, however, that with these a fan is helpful in moving warm air from the greenhouse to the home; the high roof peaks contain stratified pockets of warm air that will not readily circulate to the adjoining structure. Because both designs have glazed surfaces well above the growing areas, movable insulation that isolates the upper zone can be effectively and easily installed. The height of these two solar greenhouses can also be used to their benefit in terms of air circulation. Exhaust vents mounted near the apex will promote efficient passive summer ventilation; the added height between high and low vents increases natural exhaust circulation as is shown in the venting formula in Appendix H. Design E makes a very nice roof cover on an atrium, but we would like to caution the *novice builder* against undertaking it. The sawtooth roof must be built to handle runoff and snowloads, and in most units of this design that we've seen, roof leaks have been a problem.

Roof Design. In an attached unit the roof slope will often be predetermined by its tie-in point on your home and local snowload requirements; you won't be able to control collection by adjusting its angle of incidence. Instead, transmittance can be controlled by including an insulated/shading section that takes both summer and winter sun angles into consideration (Fig. 24).

The primary function of the insulated roof section is to keep heat losses in the warmest part of the greenhouse to a minimum in winter. Just as important, it blocks the sunlight during the summer months by reducing the effective collector area on the nearly horizontal roof surface. If the entire roof section were clear, the result would be a very large collector surface almost normal to the summer sun. The shaded and insulated section at the apex keeps the wall of the home in shadow throughout the warmest time of the year, yet provides the southern two-thirds of the greenhouse with full light.

Figure 24

The point on the rafters where the insulated roof stops and the clear roof begins can be determined for any greenhouse design by a cross-section scale drawing using the information given in *The Charts* (Appendix B). In an attached greenhouse in which the back wall is about 10 feet high and the width of the greenhouse is 8, 10, or 12 feet, the insulated portion of the roof will be the top 4 or 5 feet. With this design, the sun is allowed to strike well up the wall of the greenhouse throughout the winter period. The half-clear roof can be insulated in winter in very severe climates with blankets or rigid panels (see sliding roof panels, Chapter V, p. 48).

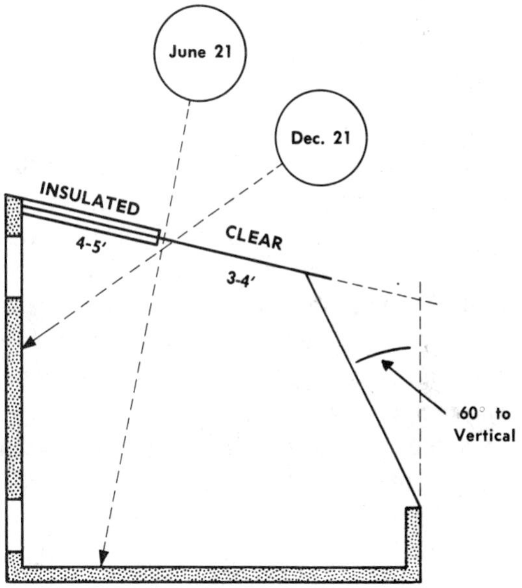

In the greenhouse shown in Fig. 25, the sun at noon will strike no higher than point A throughout the winter. You can plan to add any direct gain storage below and south of it. A wider greenhouse (E) would mean that the clear area on the roof (C) would have to be increased, creating more clear surfaces for heat loss in the winter and heat gain in the summer. Note that on June 22nd the area north of point B will be in the shade during the hottest part of the day. This means that thermal storage is out of the direct sunlight in midafternoon on summer days, when you don't need or want it heated.

The solid roof area of the greenhouse (D) will be permanently insulated. The result will be a more even temperature range in the unit throughout the year. A partial sacrifice with built-in shading is that full-light loving plants (tomatoes, cucumbers, peppers) won't do as well in the rear of the greenhouse during the summer period. However, most shade-loving houseplants, and low-light vegetables (lettuce, onions) will enjoy the covered area of the greenhouse. The loss of a little summer light is more than returned in overall thermal performance.

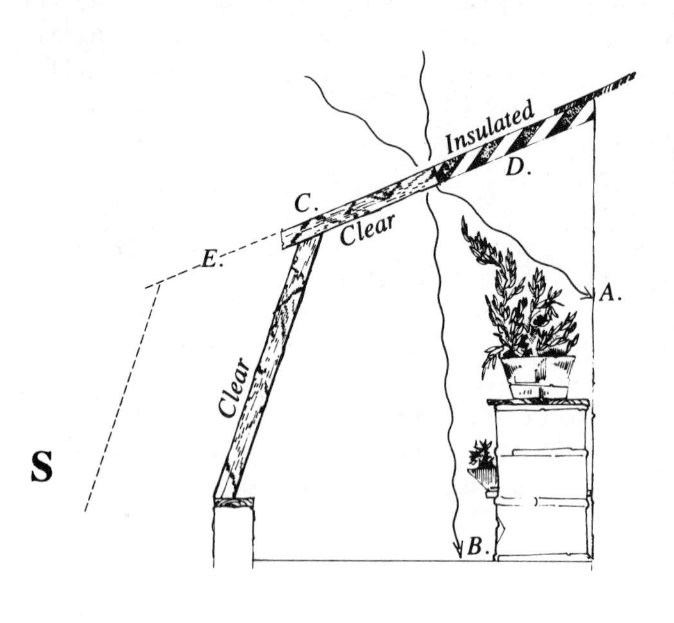

Figure 25

East and West (end) Walls. In designing the *end walls* of the greenhouse, first consider what the ultimate year-round use of the structure will be. If it is to be primarily a winter garden and used only for houseplants throughout the remainder of the year, you can make the east and west walls of the structure solid (massive or insulated frame). You'll have less heat loss through the end walls and a more efficient solar collector for supplemental home heat in the winter. However, in order for a greenhouse to have enough light throughout the entire year for adequate vegetable growth, it needs some sunlight from the east and/or west sides. South light alone is not enough. Without some side lighting, the plants will be missing the beneficial effect of longer daylight hours and blossom development will suffer. Clear or partly clear eastern walls are preferable to western exposures because the eastern glazing helps the greenhouse warm up early in the morning after a cold winter night, and the western sun in summer, combined with higher afternoon temperatures, contributes to overheating problems.

Determining the amount and location of east and west clear glazing is mainly a function of the orientation of your greenhouse. For instance, if the main collecting area (south face) is situated 30° to the east of south, then you have effectively captured the eastern (early morning) winter sunlight (Fig. 12 and 13). You should therefore make the eastern end wall solid and the western end wall at least partially clear. Remember, you're trying to get the direct sunlight from 9 A.M. to 3 P.M. in winter and about ten hours of summer sun through the clear walls of the greenhouse. As the greenhouse becomes longer on an east-west axis, the end walls shade a smaller proportion of the interior space. If the length exceeds about 25 feet and the greenhouse is under 10 feet wide, very little glazing is needed on the east or west walls.

Since part or all of the end walls will be frame or massive, they can be used for insulation, storage, or best of all, both (insulated massive). A typical insulated wall is made of 2 x 4's, or 2 x 6's, with interior fiberglass or rockwool insulation, and a polyethylene vapor barrier, then sheathed, paneled, and sealed. The advantages of such walls are that they are easy and inexpensive to build. This is the way the great majority of American homes have been built for the last four decades. The drawback with a regularly insulated frame wall is that if the heat is turned off for a time during the winter, the house gets cold very quickly. The structure is dependent upon continuous heating.

Walls with thermal mass, on the other hand, make more sense in any structure that uses direct sunlight for heat. The heat is stored in the building material and returned to the structure several hours later. Materials such as the adobe bricks used in the Southwest have the remarkable quality of delivering maximum stored heat about twelve hours after the peak collection period, when it is needed most. The old rock homes with thick walls found throughout much of the United States perform the same function. This natural cycle also works in summer, helping to keep the home cool in the day and warm at night.

For massive end walls to perform with optimal efficiency, they should be insulated on the outside. When this is done, the walls become in effect a structural "Thermos bottle," radiating most of the heat gained during the day back into the greenhouse at night.

North (adjoining) Wall. It is preferable that the north (home-adjoining) wall have high thermal mass. The heat that it absorbs and stores will be reradiated into the greenhouse and transferred through the wall into your home. The chart on p. 33 gives the thickness for *optimum* heat transfer through the wall to the

Seasonal Insulation

For those of you in extreme northern climates who wish to have both an efficient solar greenhouse in winter and a good vegetable producer in summer, consider rigid insulating panels seasonally applied to clear eastern and western walls. You can design the majority of your end walls to be clear but leave provision to fit in the insulating panels in accordance with sun movement and cold weather. Cover rigid styrofoam or polyurethane sheets with a thin sheathing and pressure-fit them between the studs on the interior side of the glazing. The panels are inserted sequentially, with the one nearest the home being installed in about November, followed by another panel as the sun rises and sets further to the south. Midwinter finds you with only south-facing glazing and R-12 side walls. The process of removal is reversed in February. This is a simple operation and is quite appropriate and cost-effective. It solves the problem of getting both summer lighting and winter solar efficiency.

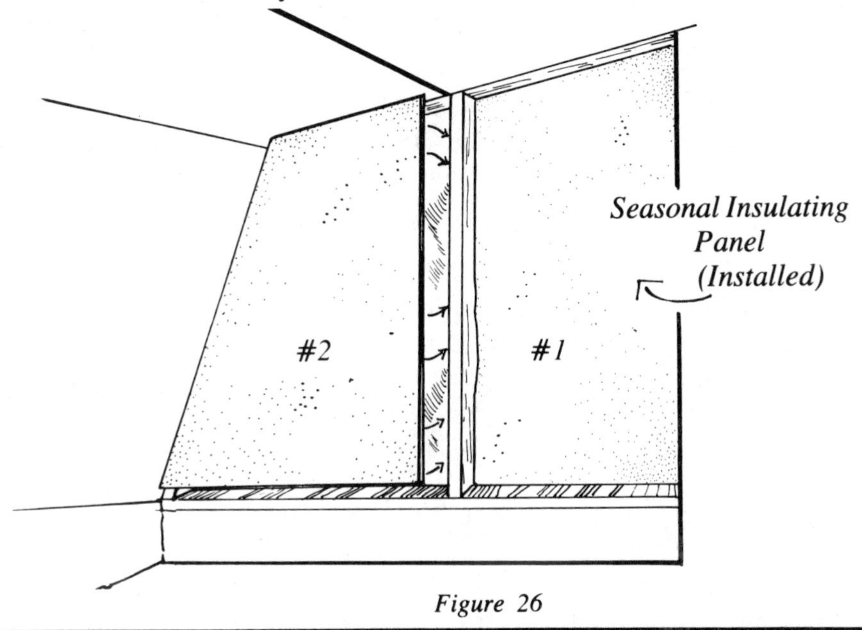

Figure 26

interior of the home. The thickness given takes into consideration the time lag between when the heat is absorbed on the greenhouse-facing side (at 100° - 120°F) of the material and when it arrives on the home side 10-14 hours later (at 68° - 78°). These dimensions can also be used in new greenhouse-home combinations.

Optimum Thickness of Massive Walls

Material	Thickness	Material	Thickness
Hollow-core block filled w/ concrete	10-12 inches	Adobe	10-12 inches
Rock	10-14 inches	Brick	10-14 inches
Water	8 or more inches	Concrete	12-16 inches

It is important to remember that the greenhouse does not need to have perfect characteristics in all of the categories discussed to be successful. However, if the unit is lacking in one area, it would be wise to compensate for this in some way. For instance, if you decide to build simple stud walls, it is advisable to increase the amount of thermal storage in the greenhouse.

Attached or Freestanding

The fundamental design choice between freestanding (independent) structures and attached greenhouses seems to need clarification. In our first edition, we emphasized the *advantages* of an attached greenhouse. They are:

1) Excess heat produced in the unit is not wasted by venting into the atmosphere, but is used in the adjacent home.

2) Cost of heating the home is thereby reduced since the greenhouse is either supplying supplemental heat or acting as an insulator against heat loss through the south wall of the home.

3) An attached unit offers a convenient location for the operator (especially when it is attached to the kitchen); the benefits of this feature will be realized most clearly during inclement or cold weather.

4) Cost of greenhouse construction is reduced since one wall—that of the home—is already built; wiring, water, and drainage hookups are handy, therefore less expensive to include if desired.

5) With an attached unit you have the option of utilizing home heat as a back-up system rather than investing in conventional or additional solar heating.

6) The owner enjoys the aesthetic and creative benefits of biosphere living more directly.

These reasons present a sound basis for deciding on an attached unit.

Another, less apparent benefit, concerns the successful operation and the general approach to attached greenhouse gardening. With the structure as an integral part of your home, you are likely to watch its progress more closely than if it were located out behind the garage somewhere. You'll be more aware of temperature fluctuations and sensitive to the progress and needs of the plants. Problems that might develop, like harmful insects, hoses left on, doors left open, can be spotted and corrected quickly.

Any style of architecture can be designed or retrofitted with a solar greenhouse. Look over the many different styles in Chapter VIII and you'll see what we mean.

Freestanding. It is possible that conditions at your site absolutely rule out an attached unit. Permanent obstructions may shade your only near-southern exposure. Perhaps your house faces south and you aren't quite ready to have a tomato patch cover your front door. Or you may simply prefer to keep the unit separate from your home. Then you want to plan an independent structure.

Here are a few advantages to an independent greenhouse:

1) The unit can be oriented exactly to true south for maximum collection.

2) It can be designed to prevent phototropic plant growth and to receive some northern light that the home will obstruct from an attached unit.

3) Your design is not limited by the configuration of the home.

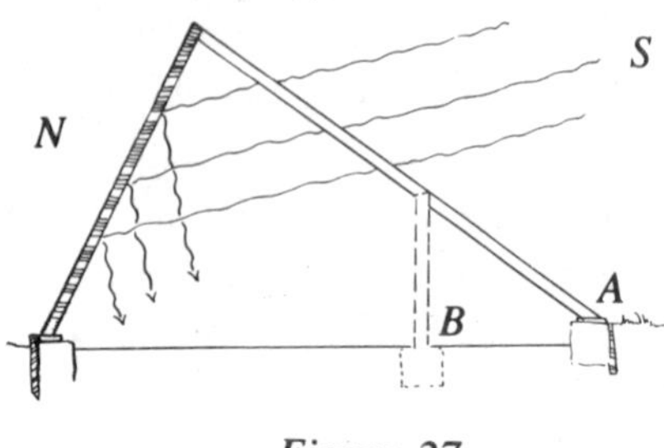

Figure 27

In an independent structure a principal design consideration is the configuration of the north wall. A common approach (developed by the Brace Institute) is to tilt the north wall toward the front of the unit; see Fig. 27. The tilt of the wall is calculated to be the *altitude* of the sun at the *summer* solstice for the latitude of a particular site. This helps prevent phototropic growth by reflecting light off of the wall to the plants. A tilted reflective north wall such as this can more than double the light the plants receive on a clear winter day. When the north wall is well insulated, this design also has about half the heat loss of a traditional independent greenhouse. The front face slopes to the ground (A) or to a low "kneewall." In areas of heavy snowfall, a higher vertical front wall is used (B) rather than extending the front face to the ground (see Herb Shop, p. 140).

A variation on this design adds a vertical north wall below the tilted, insulated, north-facing roof section (Fig. 28). The vertical wall (A) can be built using concrete or stone to provide thermal storage. Heat loss is reduced by the insulated roof and by earth-berming or sinking the structure below ground level (B).

Many innovations are found in the inexpensive, easy to build "A-Frame" greenhouse designed by Reed Maes (see p. 139). The height of a steep A-frame structure offers advantages in mounting movable insulation above plant level, as well as reflecting light down to the growing surface.

A good combination of features is pictured in Fig. 29. Here a vertical section of the north wall (A) is built with concrete for thermal mass; sloping north roof sections (B) combat phototropism. The south-facing roof section is perhaps the key to the success of the design. Like the attached greenhouse roofs discussed on p. 31, this one is partially shaded to cut down conductive/radiant losses at night and overheating during warm periods. Alternating clear (C) and solid/insulated (E) sections run the length of the roof; movable insulating panels are stored under the solid sections. The south face (D) is tilted to coincide with winter sun angles (again, note the similarity to the attached design). For a more detailed description of a functioning unit of this type, see Tyson, p. 149. Performance characteristics for these freestanding designs are also noted in Chapter VIII.

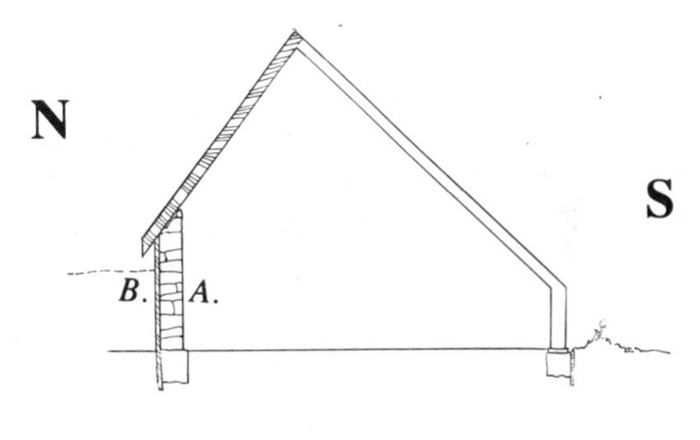

Figure 28

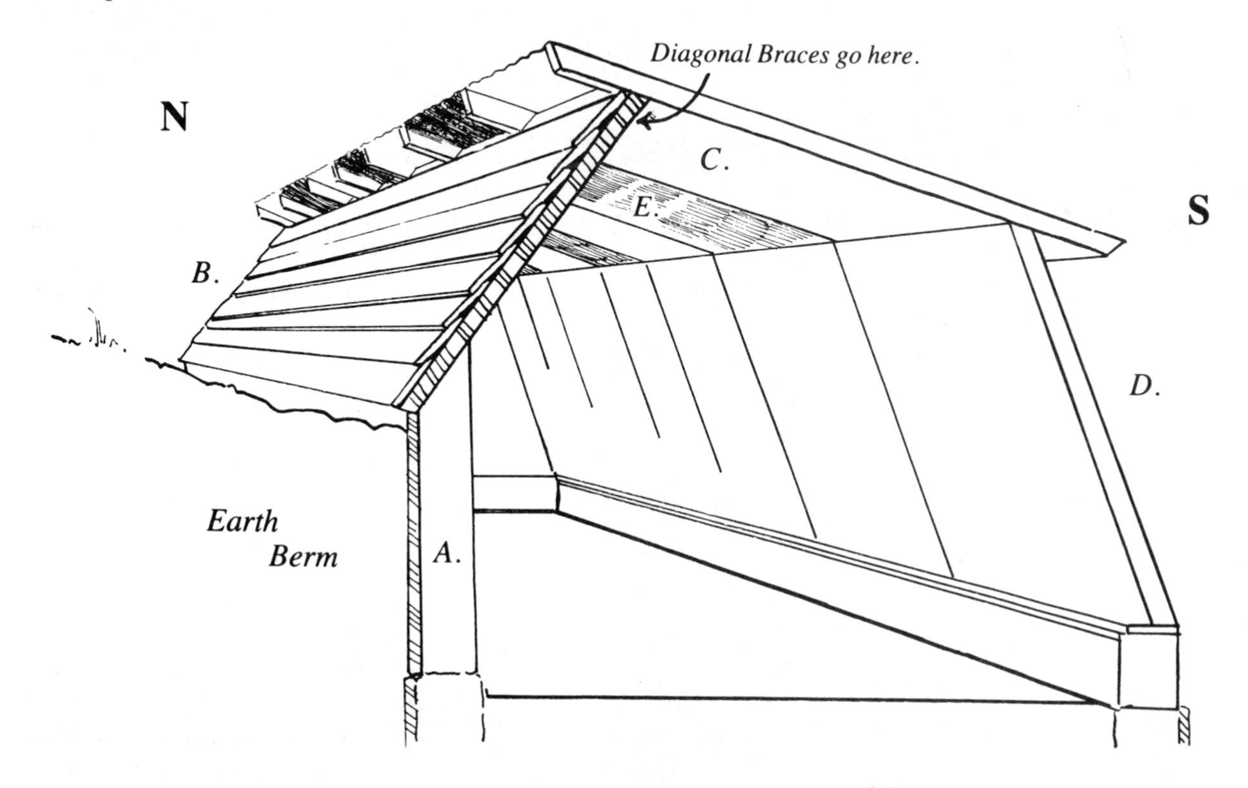

Figure 29

The design of the greenhouse interior is very personal and depends in large part on your attitude toward its use. Many people enjoy a greenhouse that provides space for activities other than gardening. If this is your feeling, allow plenty of room for sitting and moving about. You may decide to arrange the planting areas around a central living space or separate the two completely.

Most greenhouse owners, though, prefer to make maximum use of interior space for growing plants. This is a more difficult design problem that requires consideration of several important factors.

Access

If your biosphere is built against a wall having an existing doorway, the door should open away from the greenhouse area. All exterior doors are built to open out. This will allow you more freedom in arranging the interior space. It is likewise preferable to build vents that open to the outside or that slide.

Provide sufficient walking space in your floor plan for unrestricted access to all planting areas. Plants will tend to overhang the beds, so allow for growing room. At times your greenhouse may have to accommodate several people; one expanded area of walkway will furnish the needed capacity (Fig. 30).

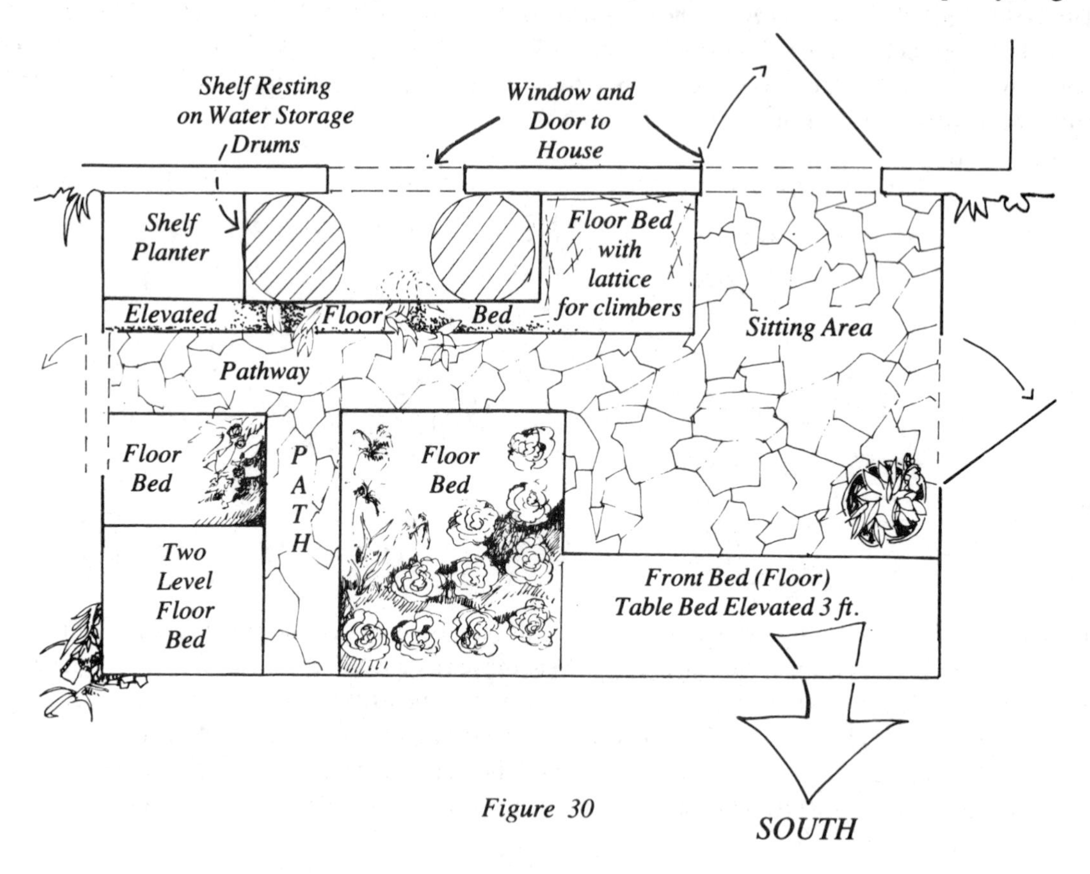

Figure 30

Planting Areas

Permanent beds may be dug directly into the greenhouse floor. If additional depth is desired, supporting sides built around ground level will hold more soil. It is important to estimate shadows that will be cast from plant-filled beds. Beds located at the rear of the greenhouse (away from the sun) may be built above ground level to prevent their being shaded by front plantings. In the small greenhouse, optimum use of vertical space is essential. This can be accomplished by adding shelves, hanging beds, planters on barrels, or by building-in tiered beds.

Table beds used in conjunction with ground beds will yield an even larger planting area. However, take care that the front planters and tables do not unintentionally block direct sunlight from those in the rear.

Thermal Storage

The use of thermal storage as a natural means of supplying heat to the greenhouse is one of the elemental principles that distinguishes a solar design from traditional ones. In this section we will look at how particular quantities and placements of thermal storage affect greenhouse temperature performance. We will emphasize *direct* (or passive) thermal storage because it is more cost-effective, readily installed, and maintenance-free than *indirect* (or isolated) thermal storage.

Direct storage is provided by various materials *in the greenhouse* that absorb heat from the sun and the air and return heat to the structure after sundown when the air (and surface temperature of other objects) drops below the temperature the storage has attained over the course of the day. It works primarily by conductive and radiant heat transfer.

Capacity. Any material has a certain *heat storage capacity* that is a function of its specific heat, mass per unit volume, and density. The chart below compares some common materials in terms of how much they weigh (their mass) and how much energy a cubic foot of the material will store if its temperature is uniformly raised 1°F.

Material	Specific Heat Constant	Density	Total Heat Storage
	BTU/lb. °F	lb./ft.^{3}	BTU/ft.^{3}/°F
Adobe	0.22	90	20
Brick	0.20	120	24
Concrete	0.23	150	34.5
Earth	0.21	95	20
Sand	0.20	110	22
Steel	0.12	490	59
Stone	0.21	165	34.6
Water	1.00	62.5	62.5
Wood	0.33	32	10.6

Conductivity. In examining storage materials it's necessary to have some practical knowledge of their *conductivity,* which determines the rate at which heat moves in and out of the material. For instance, steel is denser and has a much higher conductivity than water. A cubic foot of steel has about the same total heat storage capacity as a cubic foot (7.48 gal.) of water but will respond more quickly to temperature changes than water. If both were in a greenhouse and absorbed the same amount of energy during the day, the steel would quickly release its heat to the structure after sundown while the water would slowly give it back during the course of the night.

Thermal Momentum. Changing the sizes and shapes of thermal mass also affects the time lag of the heat released. *Fifty-five* one-gallon water containers will react differently than *one* fifty-five gallon drum. Mounted in a greenhouse, the small containers will absorb more energy per unit because they have a greater surface-to-volume ratio than the drum. But this also means that they lose the heat they have collected more quickly. Small containers of enclosed water (1, 5, and 10 gallon containers) are day-to-day thermal storage. Large containers (25, 30, 55+ gallon containers) and massive masonry walls are better for long-term storage. They can carry the unit through extended cloudy weather. It's beneficial to have both types of storage in the solar greenhouse.

Storage Media and Placement. Including direct thermal storage in your greenhouse requires added design considerations from the outset. If you plan to add a passive storage system later, allow adequate space in the original design to accommodate it. In terms of the interior space of your greenhouse, your primary concern is the proper placement of the system. Exposed water drums, for instance, should be located such that they receive maximum direct sunlight yet do not shade your plants. Thermal mass exposed to direct sunlight is *three to four times* as effective as mass that is shaded.

Fifty-five gallon oil drums full of water are a common form of "direct-gain" storage. The problem with fifty-five gallon oil drums is that they are ugly. It would take a design genius to make them otherwise. For maximum efficiency, the drums should be painted black, which doesn't improve their appearance. They are cumbersome, usually bent, and greasy. Some suggestions:

1) Try buying new, shiny, undamaged drums from the drum factory. It's highly unlikely as the oil companies that own the factories choose to peddle them full or at $25 a whack, empty. Beware of the residual contents of drums from chemical and oil companies. Some of that stuff is absolutely deadly. You shouldn't even think of having it in your greenhouse.

2) Find used drums at bread or candle factories. They have been filled with molasses or parafin and are usually in good shape.

3) Clean used barrels with gunk remover, treat them with a primer, then paint them a beautiful flat dark earth color instead of black.

Figure 31

4) Decorate the noncollecting areas of the drums to suit your fancy and taste.

5) Train plants around the sides and back of the barrels (Fig. 31).

6) Incorporate shelving and planters on top of the barrel installation. This is a good way to combine functions (Fig. 32).

7) Add a small amount of rust inhibitor or antifreeze to any metal container before sealing it to prolong its life.

Large drums should be placed near the back of the greenhouse or wherever they will catch direct sun and not shade or be shaded. If the drums are stacked on end, stagger them slightly to allow space for filling and air circulation. If you have room, lay the barrels on their sides with the filling holes facing up. See that the drums do not touch the greenhouse walls or they will conduct heat to the walls that could be used to raise ambient air temperatures.

However ugly they might be, the effect of water storage drums on your greenhouse can be beautiful. If direct sunlight raises the temperature

Figure 32

of the water in a fifty-gallon drum thirty degrees, you will have stored about fourteen thousand BTU's of heat energy. Water stores *three to four times* as much energy per pound as an equivalent amount of rocks and masonry.

A variety of other water storage containers can be employed. They include water-filled beer cans, glass and plastic containers of all sizes (Fig. 33), discarded gas tanks from cars and trucks, rubber and vinyl pillows, open metal vats and galvanized steel culverts. The important thing is to check that no direct-gain storage unintentionally shades storage behind it. For this reason various-sized containers, lower to the south and higher to the north, are advisable.

If you can't or don't want to use water containers in your greenhouse, other mass, such as masonry walls, planters made of stone, and the earth in ground beds can be functional and beautiful thermal storage. A six-inch concrete slab and/or brick added to the floor can act as a "heat sink" for storing thermal energy.

The advantages of using these materials are that they don't have to be contained like water and they all have different time-lag characteristics. By combining various materials with differing thermal properties, the whole biosphere benefits, aesthetically as well as practically. Of course, all thermal storage materials should be a dark color (or black) for efficient absorption of energy.

Sizing Mass

The following charts show approximate temperature fluctuations in a solar greenhouse in relation to

Figure 33

the amount of thermal mass within the unit. The thermal mass is sized to *one square foot* of south-facing glazing in three different amounts to demonstrate various operating modes. The performance estimates assume that the majority of the mass is visible to the sun and that the design of the greenhouse is similar to the model used in this book. No movable insulation is applied in any of these cases.

Low Thermal Mass. As indicated by Chart I, this operating mode results in high temperature fluctuations. This is what will happen to conventional greenhouses if they are left unheated in winter. Clear-day temperatures will exceed 100°F while nighttime temperatures will drop down close to the outside air temperature.

There are several operating modes where little or no thermal storage may be justified. One would be an extremely cold and cloudy climate like parts of Wisconsin and Minnesota. The attached greenhouse might be intentionally designed to extend the summer growing season for very little cost and with minimal construction; for instance, it could be as simple as a framework of 2 x 4's tacked onto the south side of the house with polyethylene stretched over the frame—maybe $100 investment. The greenhouse would still be a good solar collector through the winter and deliver perhaps 700-900 BTU's of heat per square foot of glazing to the adjoining house on clear days. In practical terms, this would be extremely low-cost heat delivery. Of course, the plants will freeze by early November and the unit won't provide good growing conditions again until March.

A variation of this mode has been applied in several new homes and buildings in severe climates. The designers build-in limited thermal mass in order to raise ambient air temperatures; the surplus heat is then ducted into isolated rock storage beneath the floor of the new home (Fig. 34). In this design isolated thermal storage can be effective because the reduced thermal mass in the greenhouse allows ambient air temperatures to get high enough to really charge the rock bed.

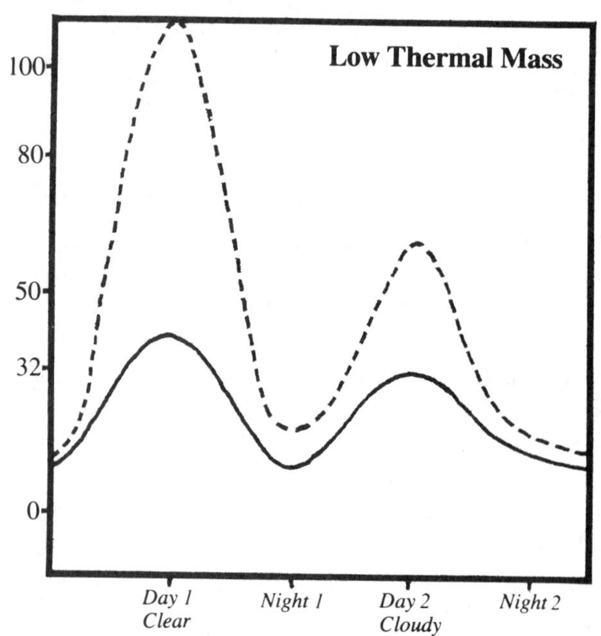

Chart I

The deep South and Gulf Coast region are other areas of the country in which greenhouses need little or no thermal storage. These areas generally have less than 2000 degree days, so not much thermal mass is needed to maintain adequate growing temperatures in the greenhouse. This makes the unit easier and cheaper to build. The greenhouse should, however, have at least one gallon of water per square foot of glazing to maintain temperatures above 50° in winter. The surplus heat (there will be an abundance of it) should be moved by fan directly to the house or to isolated thermal storage. Be prepared also to use a fan to cool the greenhouse spring through fall.

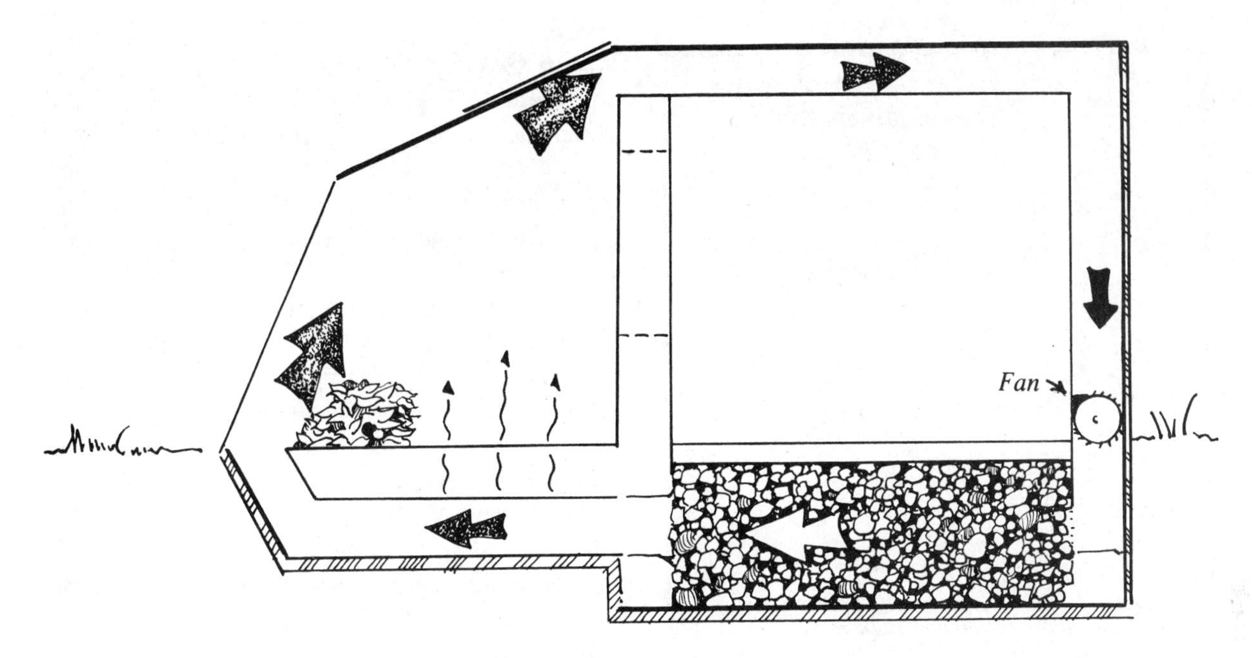

Figure 34

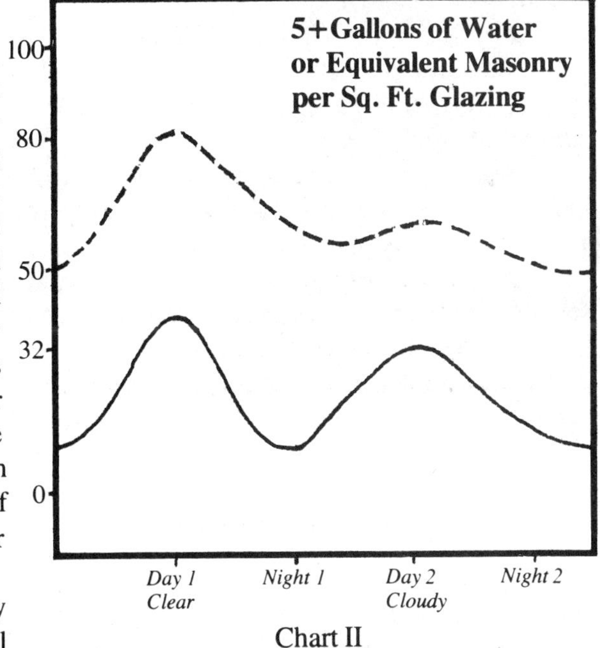

Chart II

Five or more gallons of water or masonry equivalent. From Chart II it is apparent that increased thermal mass greatly decreases the air temperature fluctuations. The mass acts like a heat "sponge," soaking up heat during the day and releasing it back to the greenhouse at night and during cloudy weather. This mass is recommended for independent solar greenhouses and for attached units in which very cold-sensitive plants (usually ornamentals) are being grown. Increasing the mass has the effect of giving the greenhouse longer "staying power" in cloudy climates. The storage capability of the massive material in this quantity can carry a greenhouse through three to five days of very cold and cloudy weather without interior temperatures dropping below freezing.

At first glance it would seem that there is no way we could go wrong by adding more and more thermal mass until the temperatures stabilize between 55-75°. However, it doesn't work that way. A solar greenhouse in cold climates (4000–6000 degree days) added to a frame home can be *overmassed*. If the unit is being used for supplemental home heating, the extra mass has the effect of robbing heat from the greenhouse air that could be used in the home.

In warm humid climates a different problem exists. In regions that don't experience a consistent *diurnal* (day to night) temperature swing, such as the Southeast and Gulf Coast, this mode can cause overheating problems in the summer. If the nighttime temperatures *don't drop* below 68-70° for long periods of time, the thermal mass has no way of cooling down. The result is that the thermal mass slowly increases in temperature throughout the summer and before you know it the enclosed water storage is 90° on an August morning. This isn't good for the plants or the adjoining home.

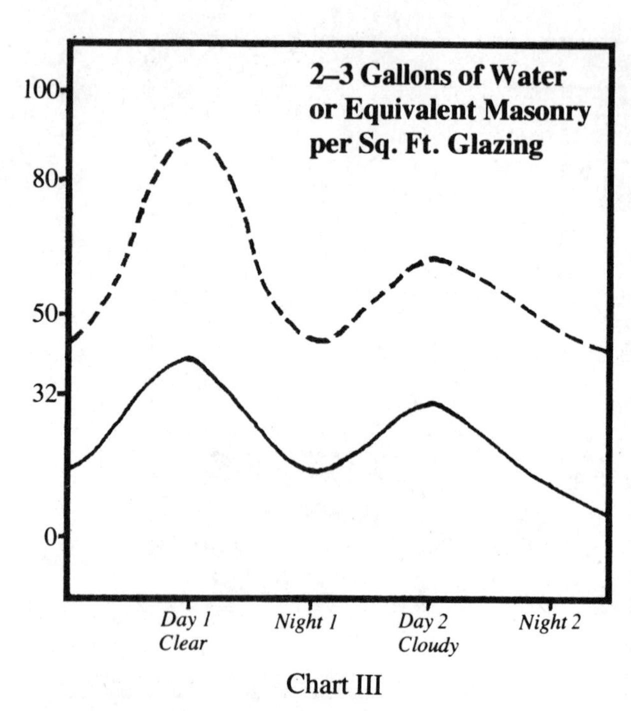

Chart III

Two to three gallons of water or masonry equivalent. This is the quantity of mass we recommend for the majority of attached solar greenhouses. With this amount, the unit will maintain temperatures 25-35° *above* the outdoor lows in winter. The storage can carry the unit through about three days of heavy clouds and cold weather (down to zero) before freezing becomes a danger. In 6000+ degree-day climates with 50%+ solar conditions it is likely that supplementary heat and/or movable insulation (p. 44-48) will be needed a couple of times in the winter. The greenhouse will maintain good gardening temperatures from late February through mid-November. With less mass to absorb heat than in the preceding operating mode, the ambient air temperature in the greenhouse will be higher on clear days, and the surplus heat can be shared by the adjoining structure.

Isolated (Indirect) Thermal Storage. Increased heating performance can be obtained from a solar greenhouse when excess heat in the apex is tapped and moved by a fan system to isolated rock beds or water storage. This extra storage component can be located either in the greenhouse or below the floor of the adjoining home.

There are three advantages to such a system: 1) daytime temperature peaks in the greenhouse are lowered; 2) heat that would otherwise be wasted is moved to an area of the greenhouse or adjoining home where it can be used; and 3) heat storage is increased in the greenhouse or home.

The drawbacks to isolated thermal storage are increased cost of installation, occasional difficulty in servicing and maintenance, and a dependence on outside power sources (usually electricity) to drive the fans or pumps that move the heat.

Passive Solar Associates of Santa Fe, New Mexico, have generated some practical advice about planning isolated storage used in conjunction with solar greenhouses.

- Warm air from the greenhouse is introduced into a north plenum and is pulled or pushed through 1½-2½'' pebbles to a south plenum and back into the greenhouse.
- Rock beds charged with heated air from a working greenhouse use low-grade heat from 70°-90°F. Since they require larger fans, ductwork, and controls to distribute the stored heat, radiant or conductive distribution is suggested.
- A concrete slab can cap (or top) a rock bed and become the floor of the home or greenhouse. From the rocks the passive heat is transfered to the interior space up through the massive radiant floor. (See Unit 1, First Village, p. 119.)
- If a frame floor covers the rock bed, the storage should allow vertical air flow. Heat enters the top of the rock bed and is forced through the *bottom* of the plenum back to the greenhouse. The heated rocks provide convective heat through the wood floor.

- Rock beds must be well insulated on the sides and particularly at the bottom. Rigid insulation is advised with a six-mil. vapor barrier below the low air channel. Feed air into the corner opposite from where it is taken out for even distribution through the bed.

Some other rules of thumb for planning a rock bed are:

1) Use a sufficient air flow to move the required heat at the low-temperature differences available.

2) Use a high-flow squirrel-cage fan to obtain high efficiency and quiet operation.

3) Complete the air flow circuit by returning the air to the greenhouse.

4) Don't figure on using more than 1/3 of the net heat out of the greenhouse.

Since accurately sizing a rock bed is contingent on pressure drop, face area configuration of the bed, face velocity and rock sizes, it becomes more complicated than this presentation suggests. If you plan isolated storage use the rules of thumb given here as *guidelines* for discussion with a competent engineer or designer.

Complicated high-temperature hot-water systems cannot presently be economically justified in greenhouses. Avoid them. Use air or low-temperature water as the heat-exchange medium if an active system is what you need. Generally we have found that *passive* storage arrangements that include a massive north wall and well-distributed thermal mass throughout the greenhouse interior are sufficient to meet most climatic situations.

Convection and Ventilation

A well-designed solar greenhouse has a passive convection cycle established by proper venting to the adjoining structure. Heated air in the greenhouse rises and flows through a high opening to the home. A low opening in the shared wall allows cool air from the house to enter the greenhouse for heating. Without any mechanical devices this natural cycle will function continuously on any relatively sunny day. The plants in the unit convert carbon dioxide into oxygen-rich air for the home, a definite health benefit for the occupants. Also, in areas of the country with very low humidity, the added moisture from the greenhouse will be welcomed in the home.

The rate at which daytime convection occurs determines how much heated air can be delivered to the house. The faster the air moves the more heat is being taken from the greenhouse. It's important that the air movement rate is correct so the greenhouse stays warm enough and still provides usable heat for the adjoining structure. The rate at which air moves is measured in *cubic feet per minute* (cfm). For an acceptable temperature range in the greenhouse and usable heat for the home, we recommend an optimum rate of 4-6 cfm per square foot of south-facing glazing. So in an attached solar greenhouse with 200 square feet of south glazing, the air should be moved to the home at a minimum of 800 cubic feet per minute. In a unit 16 feet long, the total volume of greenhouse air would be exchanged with the home about every two and one-half minutes. If the air were moved at a faster rate, say two air exchanges from greenhouse to house per minute, the result might be a greenhouse running at 60°F and blowing 60° air into the house, which wouldn't help either structure.

Vents must be properly spaced in relationship to one another. When high vents are located directly above low ones, the air circulation pattern becomes somewhat localized rather than diffused over a wide area, and the majority of the airstream will travel the shortest distance and not spread its heat well throughout the adjoining room. For better circulation, get the air moving over a greater distance by staggering the vents (Fig. 35).

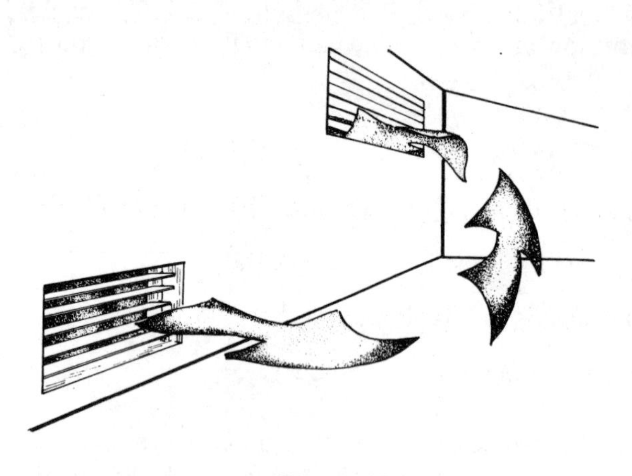

Figure 35

You can also move the air with a small fan rated to move this volume of air (or slightly more to make up for friction losses). It can be purchased through a catalog fan supplier at a low cost.

However, many people prefer not to use any active machinery in their greenhouses but simply let the sun do all the work through passive convection.

The greenhouse must also be vented to the outdoors for summer cooling. The openings should be located to promote cross ventilation, preferably one in each end wall (or an end wall/roof combination). As with the openings to the home, it is important to have both a high and a low exterior vent. Place them as high and as low as your design permits; all venting systems (exterior and interior) operate best with the maximum vertical distance between them.

We offer the following general recommendations for vent sizing:

When using	*to*	*allow*	*of*
Windows and doors	heat home	10-15%	wall area covered by greenhouse
High and low vents	heat home	6-8%	wall area covered by greenhouse
High and low vents and exterior door	exhaust greenhouse in summer	15-20%	floor area of greenhouse (rate door at 1/2 actual square footage)

For formulas to calculate exact vent dimensions for any greenhouse, see Appendix H, p. 189.

Movable Insulation

In a well-sealed greenhouse, the greatest losses will occur through the clear wall and clear roof areas. Traditional (all clear, exposed) greenhouses rank among the most energy inefficient structures ever built. That is why solar greenhouse designs use clear surfaces only where absolutely necessary for heat collection and plant growth. Still, energy will be lost at an amazing rate through the glazing when the sun goes down. The best way to reduce these losses is by installing movable insulation. With the benefit of this added insulation, the greenhouse becomes more efficient in storing energy overnight and during extended periods of cold weather. Because the average nighttime temperatures are higher, the greenhouse warms up more quickly in the morning. The warmer, more even temperatures provide a better growing environment for young plants, increasing both the rate of growth and production. The greenhouse has a longer "staying" period, that is, it is not as dependent on a steady dose of sunlight to recharge it and maintain growing temperatures. The unit is also able to provide more heat to an adjoining structure since it does not lose as much to the outside. Movable insulation can eliminate the need for an active solar heating system, increased storage, or conventional supplemental heat.

In many parts of the country where the winters are very cold and where the sun does not shine with great regularity (Great Lakes area, New England), movable insulation is a necessity. However, several

If the openings between the house and greenhouse are left open in the evening, the convection cycle will be reversed. This is called *reverse thermosiphoning,* and its effect is to take heat from the house and move it to the greenhouse. Reverse thermosiphoning might be intentionally planned for extremely cold nights when you want to supplement the internal passive heating of the greenhouse. Generally, however, this pattern of air flow is not desired or needed. One way to stop it is to close the greenhouse off completely from the house; if windows and doors are the interior venting arrangement, then it is no problem to shut them after sundown, and that stops reverse thermosiphoning immediately. But if you have built high and low vents, it might be difficult to reach them, and in this case you may want to install an automatic *check valve*. Well, here it is, the 3¢ solution, suggested by Doug Kelbaugh, who applied the idea to his Princeton, New Jersey, home.

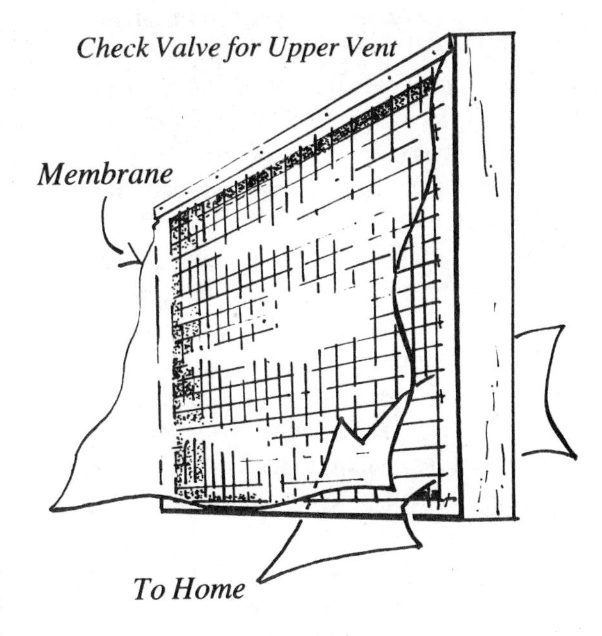

Figure 36

This is a failsafe check valve. Applied to the top vent, the lightweight plastic film (or silk) is placed on the house side of the wire mesh. On the lower vent the plastic film is on the greenhouse side. When these check valves are in operation, the greenhouse can only supply heat to the house, never take warm air from it, because the film is forced back against the screen by the natural convection currents and stops reverse thermosiphoning.

factors have to be taken into consideration in choosing the type that is best suited to your greenhouse (and your budget). You will have to plan for the appropriate system in your interior design.

1) Most systems are bulky and thus can be difficult to store, move, and remove. Interior rigid insulation panels have to cover large glazed surfaces, and the problem is to provide a storage space for them when they're not in use. Flexible curtains or drapes are not much easier to store because, in order to achieve a respectable R-factor (insulation value), you have to use a good insulator, such as two inches or more of fiberfill. Bunching a 10 x 20-foot blanket that is two inches thick in the corner of a small greenhouse can make The Blob look handsome by comparison. Either an adequate space has to be designed-in to the greenhouse for storing the movable insulation or space has to be provided in a room, garage, or tool shed.

Exterior rigid panels are exposed to the weather and thus have to be protected by some durable material. They are often sheathed in plywood, which is effective but makes the panels heavy and so increases the awkwardness of moving them. Various design innovations have been applied to overcome these problems, and we will discuss them shortly.

2) Some systems are expensive, costing up to $14.00 per square foot. Although such systems are very ingenious, demand little or no effort on your part to operate, and solve the storage problem, it may be difficult for you to justify paying this much to insulate the glazed surfaces when the greenhouse only costs $7.00 per square foot to build.

3) The success of a manually operated movable insulation system depends a great deal on the greenhouse owner. You have to be present to install and remove the panels daily during cold weather. If you miss a night and the temperature drops unexpectedly low, you can lose your entire greenhouse garden. So, while manual systems are generally more economical than automatic ones, they do require a daily time commitment.

4) Some materials can pose a health hazard; polystyrene and urethane, for instance, will give off toxic fumes if a fire occurs, and city building inspectors have told me that they will not approve these materials for use without a fire-retardant covering on all surfaces. Although I have never heard of any problems of this nature in a greenhouse, it is best to be aware of the potential hazard of using certain products without the appropriate protection.

With these various considerations in mind, let's look at several different systems and examine their advantages and disadvantages.

A time-tested device that can insulate your clear walls is a curtain or drapery (Fig. 37). Pulled closed in the evenings or on cloudy days, the curtain should seal as tightly as possible to be effective. Ideally the curtain should have a reflective surface (such as aluminized Mylar or Foylon) facing the air space in the interior. A disadvantage to the curtain is that you have to be present to operate it (unless, of course, you install a motor with a heat- or light-activated switch).

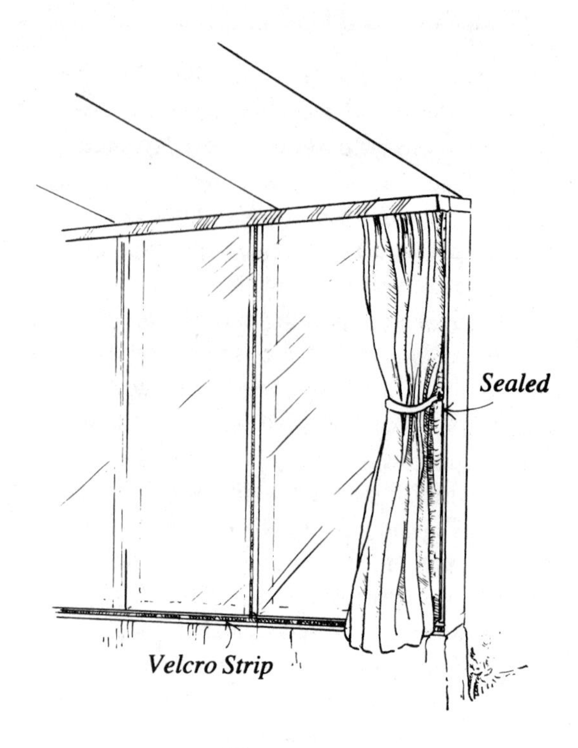

Figure 37

Another approach is to use rigid foam panels to fill clear areas at night (Fig. 38). Clips and magnets have been used to attach the rigid panels (see Zomeworks, p. 161, Chapter VIII). The obvious disadvantage is that the panels must be removed and stored during collection periods. You might devise a mounting system for the stored panels in which they double as interior reflecting surfaces to combat phototropic plant growth.

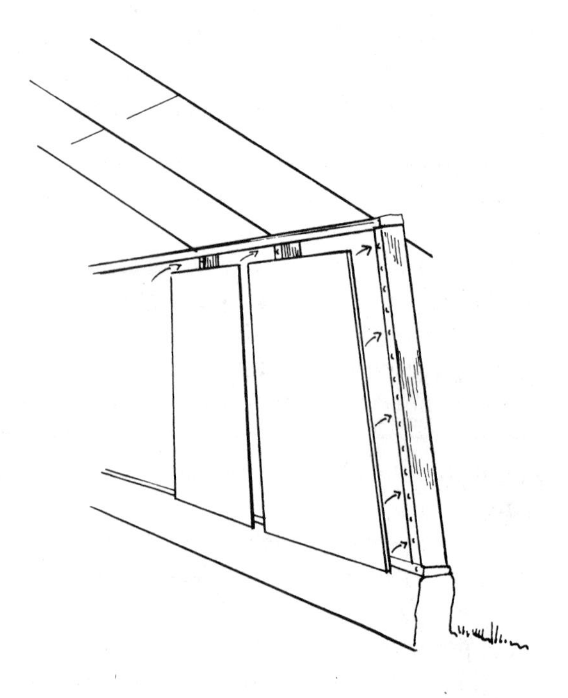

Figure 38

An insulating curtain can be made from rigid styrofoam or polyurethane panels sandwiched together with cloth on alternate edges (Fig. 39). This oriental style curtain is stored at the top of the clear wall and lowered via cord or wire that runs through the panel centers from top to bottom.

In any vertical curtain arrangements, the fit must be tight when closed or the system may defeat itself due to the "chimney effect." Thermal action can set up convection currents between a loose curtain and the clear wall, causing heat loss greater than if the curtain were not there.

The rolling blanket shown below (Fig. 40) has been used in storm window applications and can be adapted to your greenhouse if both layers of glazing are mounted on the outside of the front face. Tracks on each side and a Velcro sealing strip at the bottom prevent the chimney effect. There are several companies manufacturing insulating blakets using this scheme. They are featured on pp. 159-160.

Insulating materials on the outside of the structure offer another solution to the heat loss problem. The one shown in Fig. 41 that doubles as a reflector is an example. We suggest that you investigate buying manufactured panels that have lightweight aluminum facing adhered to the rigid insulating foam if you try this installation. Insulation distributors or mobile home manufacturers might give you a lead.

Clear roof sections can be insulated in several ways. The first option, developed by Scott Morris, (Fig. 42) employs rigid insulation that is stored under the solid roof during collection periods. The panels slide down to cover the clear roof at night. They are also effective in shading the greenhouse during the summer. This insulation system must be planned before the roof is built, as it requires specific materials and design considerations.

You can also install a hinged insulating panel that swings from its storage position under the solid roof to cover the collecting surface (Fig. 43).

One of the most cost-effective methods of reducing radiant heat loss through the roof *and* front face is to isolate both areas from the growing space with a movable blanket suspended from wires

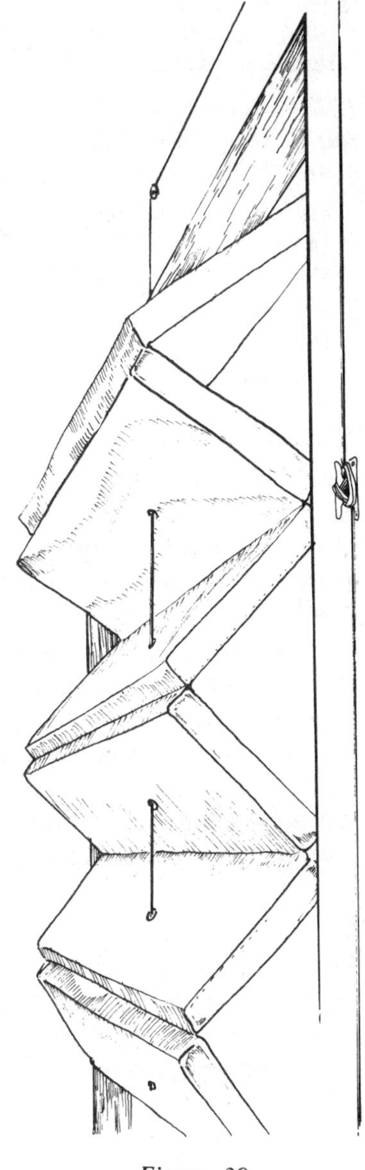

Figure 39

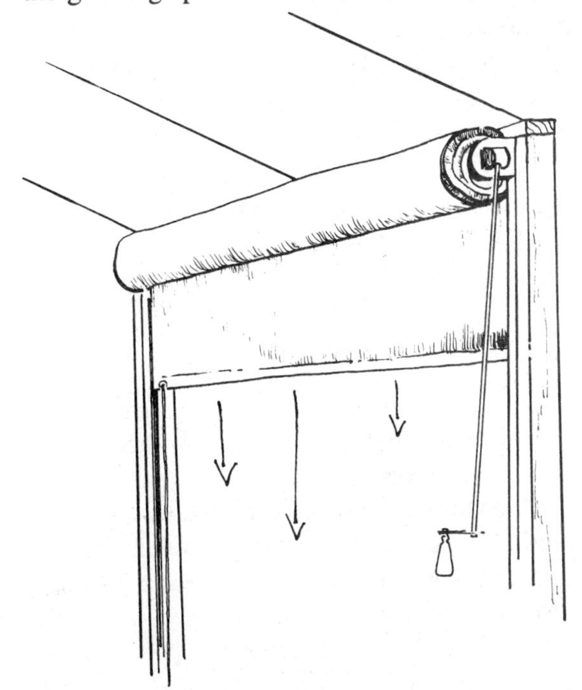

Figure 40

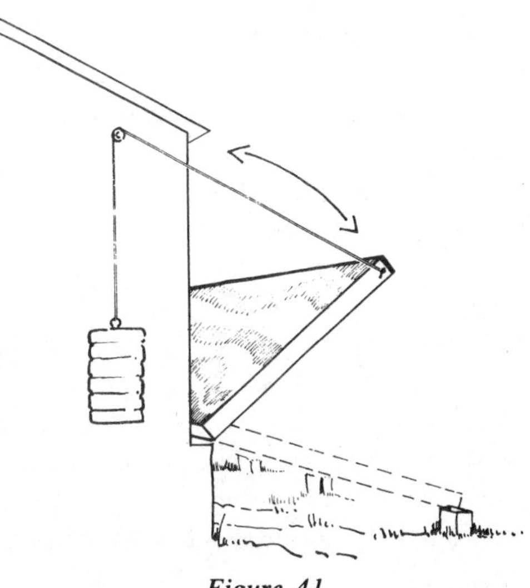

Figure 41

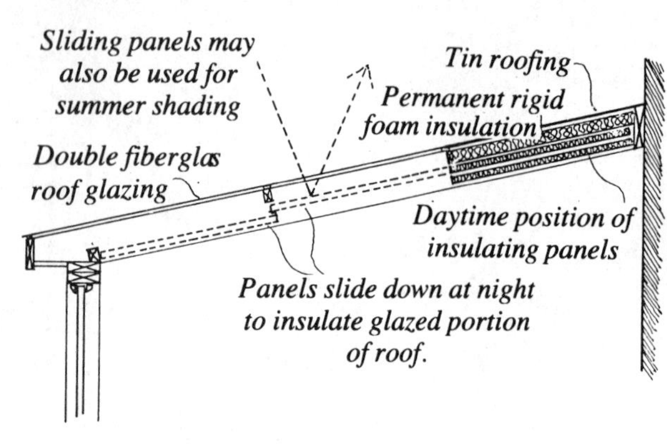

Figure 42

(Fig. 44). Inexpensive shade-cloth materials made by several commercial greenhouse supply companies are effective for this installation. These large coverings, originally marketed to keep light out of greenhouses, can serve a dual function by keeping heat inside the unit. They are usually a very thin woven cloth with virtually no R-value. The material has one reflective side that should face the interior. When mounted to block all of the glazed areas, the shiny side bounces about 90% of the radiant energy back down to the plants and the thermal storage below. If properly installed, the blanket can also act to break up convective loops and slow conduction losses through the glazing. Tests have shown that a simple one-layer 1-R reflective barrier can reduce heating needs in commercial greenhouses by up to 60%.

We recommend radiant barriers before higher R-value insulating blankets. The low-R barrier can serve as your primary clear-surface insulation in all but the coldest climates. A combination of the radiant barrier and insulating blanket or panel will probably be necessary in extremely cold regions if a full winter garden is to be maintained.

Several universities and chemical companies are experimenting with a poly-film application that is opaque to the infrared rays. One such product is Heat Mirror, developed by Day Charoudi and John Brooks (Suntech Research Associates, Corte Madera, California). A selective surface, vacuum-coated with transparent mylar, allows 70% transmission of solar radiation and a clear view through windows on which it is mounted. The coating cuts down reradiation of long waves by 75%. Acting as an effective radiation trap, the Heat Mirror increases the thermal resistance of double-glazed surfaces 3½ times. In greenhouse applications this type of product could be used within clear roof sections as an insulator. The 30% decrease in incoming solar radiation produced by the film could be compensated for by enlarging the clear roof area by about one third.

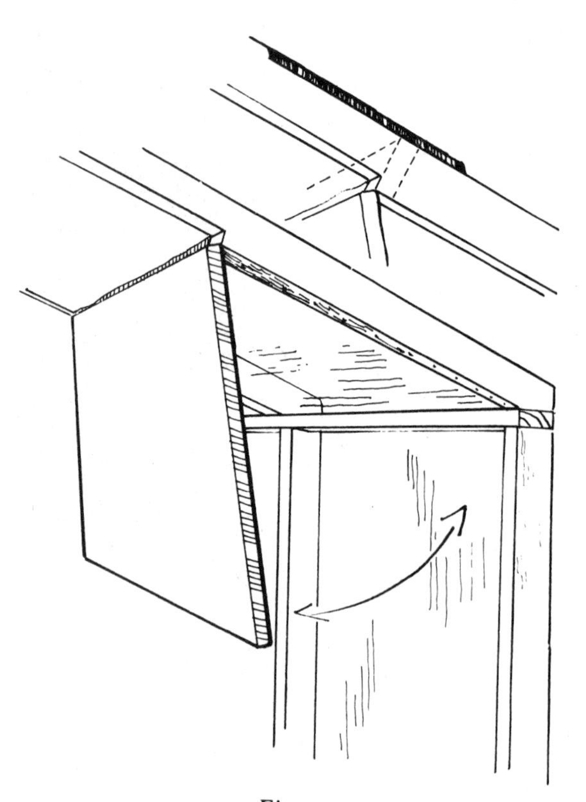

Figure 43

Another "automatic" insulating system has been developed by Zomeworks in Albuquerque, New Mexico. The design is active in that it uses electrical blowers to fill a double-glass clear wall with small styrofoam beads when temperatures drop (Fig. 45). The "Beadwall"® acts as a solid insulated barrier when filled and a transmitting clear wall when the beads are automatically vacuumed out.

Aesthetics

The main goal in designing your greenhouse interior is to produce an environment that is both practical for growing plants and beautiful to experience (for you and the plants). Your choice of an interior layout will to an extent be predetermined by functional considerations. Many designers equate function with beauty—the most functional is the most beautiful. Since the function and process of the greenhouse is such a positive, creative one, this

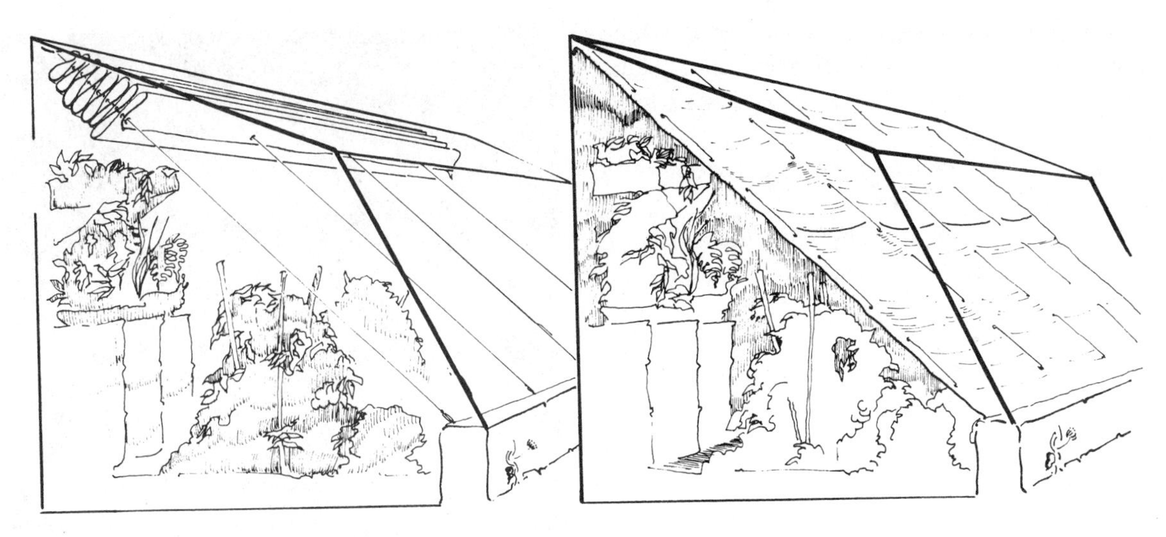

Figure 44

notion will generally hold true. As you concentrate on producing the most functional space for growth, certain characteristics that are unique to the biosphere environment will guide you in deciding how the interior should look.

You may refer to Chapter VII for planting information to determine the most suitable growing arrangement for your purposes. Certain depths of soil and conditions of light and shade are recommended for different varieties of plants. These conditions will influence the dimensions and placement of beds and interior structures.

Growing, changing plants will be a primary part of your greenhouse. You may choose to heighten the profusion by using a wide variety of colors, shapes, and surfaces in the walls and bed structures. On the other hand, you may wish to balance the active look of plant life with a more simple interior design by limiting the number of exposed (visible) materials to a few that appeal to you most. Choosing few colors will accomplish the same end.

If a constantly changing environment suits you best, build trellises for growing "living walls" within the space. Climbing vegetables such as tomatoes or pole beans will cover a hanging string or stake lattice in short order. The planting will necessarily demand consideration of shading, but the aesthetic rewards can be great.

The finishing touches you apply are a matter of personal taste. Surfaces that receive direct sunlight and have sufficient thermal mass for heat storage should be painted with a dark nonreflective color. Flat black is most effective for absorption, but we

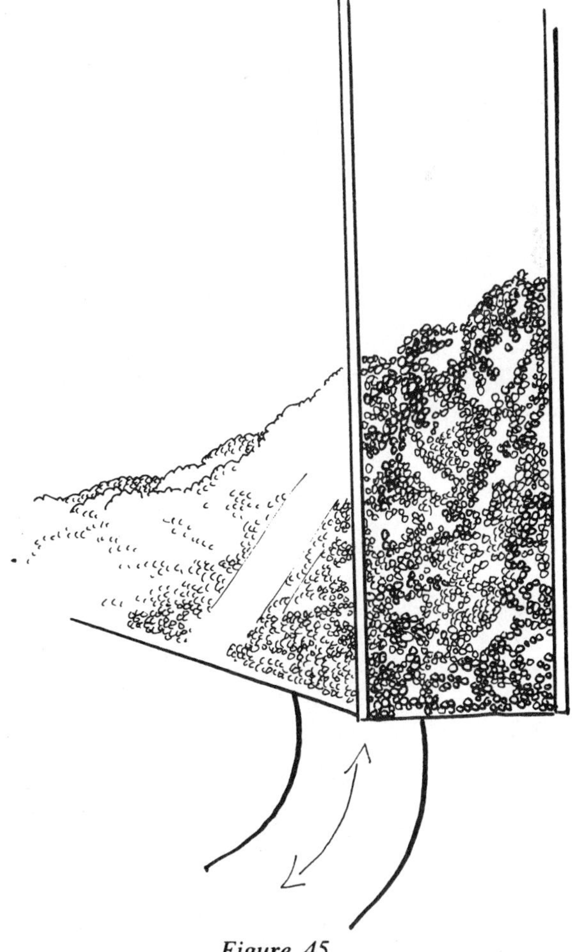

Figure 45

also recommend a dark blue, brown, red, or green. All insulating walls with no thermal storage potential should be a light color to reflect more light to the plants. Trim, such as wood stripping, molding, or lath can be used attractively to unify planes (and to cover less-than-tight joints).

Perhaps the single most important step in designing the greenhouse interior is to examine your own likes and dislikes. The qualities of environment that you feel are most attractive should be combined in the plan. The value of this point has been proven repeatedly: if you are pleased with the final results, you will spend many pleasant hours in the environment, helping to insure its success and deriving the fullest benefits from the fruits of your labor.

Figure 46

CHAPTER VI CONSTRUCTION

In this chapter we will take you from the first shovel of excavated earth to the last nail in the wall. Experienced builders may want to skim this material for pertinent greenhouse information and do the rest of the building their own way. In this edition we have included several options for certain of the construction steps. They will allow you to better plan for your specific site, geographical region, and budget. Also, the construction techniques described here would apply to any basic add-on green house geometry. A list of tools and materials used in building a 10 x 16 foot attached solar greenhouse is shown at the end of the chapter. But read the entire chapter before you buy or build anything.

There are about a thousand ways to do any particular building operation (that's what perpetuates the mystical aura that surrounds construction). Any ''expert'' will give you one or two of these ways. So will we.

Once you start building, you'll find it addictive. We have had a love-hate relationship with it for years. It will torture your days and keep you awake nights, but you'll want to do more. You will think of all the improvements you could make in your house and you'll be hooked.

General Tips

1. Plan each step of construction as completely as possible. Determine which materials you need for each step and buy them before beginning. This is easy to say and not so easy to do. But if you consider the amount of time, organization and gas money that goes into a ''quick trip to the supply store,'' you should be convinced of the need to plan ahead.

2. Many construction materials come in standard sizes. Sheetrock and plywood, for instance, come in 4' x 8' panels. Designing the dimensions of the greenhouse to correspond to these standard sizes can eliminate time-consuming cutting and custom fitting. It also reduces expensive waste.

Do not expect framing lumber to be the size it's called. Those days are long gone. For instance: a 2x4 is 1-1/2''x3-1/2'', a 1x6 is 3/4''x5-1/2''. Also, when buying stud lumber, check the length. Not long ago, I bought some 8'' studs that shrank to 7 - 7-1/2'' by the time I got home. When I called the company, I was told that including the top and bottom plates, the studs would make an 8 foot wall. Well, that's true, but I paid for eight feet!

3. Prices vary greatly. In shopping for materials, a few phone calls to competitive suppliers may provide substantial savings. If you can get some other folks interested in building a greenhouse, you can save some money by making ''quantity orders.''

Americans don't realize that there is *no set price* on anything. Don't be afraid to bargain-shop or haggle. Ask for a special price. Get to know the manager of the store. Tell that person that you're working on an ''experimental project.'' That always sounds interesting. It may bring the store increased business if your friends like your greenhouse and decide to build one themselves. Supply and demand still functions in the building industry. If an item isn't moving, the manager can lower the price on the spot.

4. You may choose to use less than ''first grade'' materials. Resawn or rough wood can be purchased for about one-third the cost of finished grade lumber. Usable materials are often discarded; you can recycle them. Check large construction sites, salvage yards and dumps. A note here: How much recycled material you use may depend on how much time you have to devote to this project. The reason most construction companies buy everything new and in standard sizes is to save time. If you have the time, save cash by scrounging.

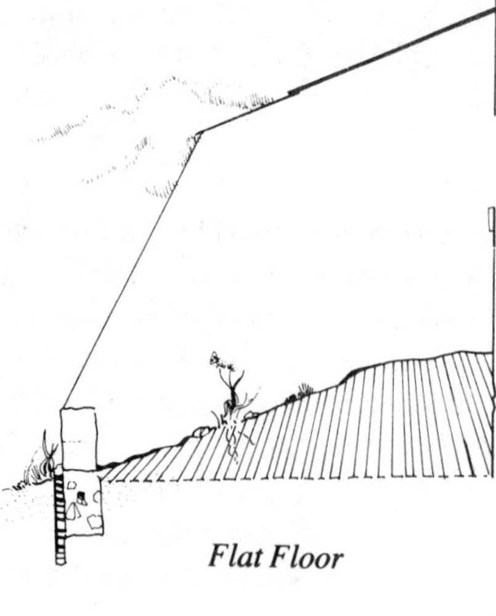

Flat Floor

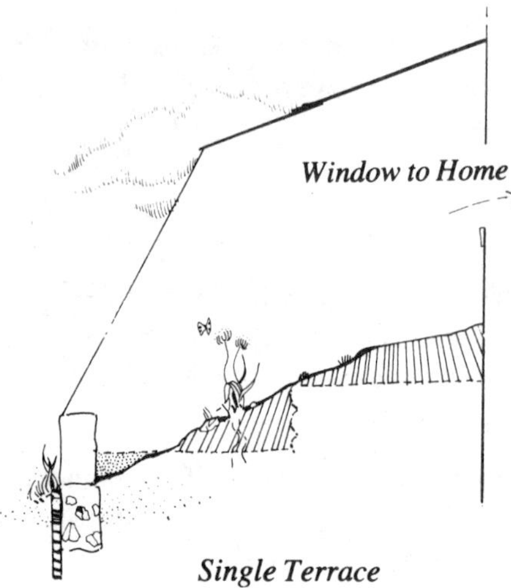

Single Terrace

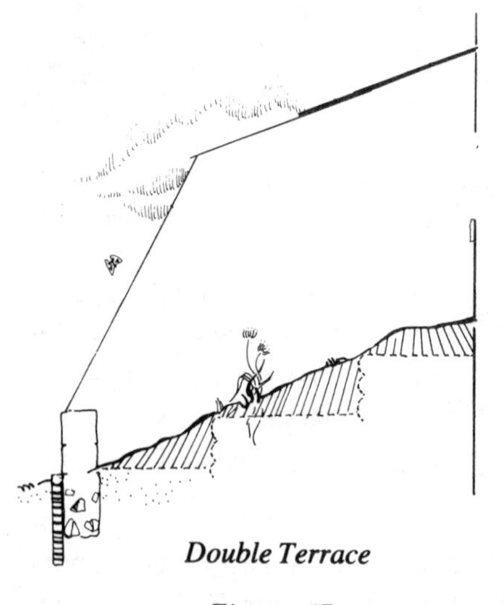

Double Terrace

Figure 47

5. In buying materials, order somewhat more than you expect to use. This will allow for mistakes and save unnecessary trips to the supplier. It's also wise to expect the total cost of construction to be somewhat higher than your estimate and time involved longer. (I'm usually off by about twenty percent.) This is probably why the building industry has the highest percentage of new company failures of any industry in the country. When was the last time you heard of a construction project being completed in less time and at a lower cost than the estimate?

6. Set up a staging and storage area for the materials. Plan to keep the entire building area off-limits to little children and pets. There are so many activities going on at a construction site it's easy for someone to get hurt.

7. In all steps of construction, measure as accurately as possible. If in doubt, exceed the correct measurement rather than cutting under it. You can always take a bit more off but it's difficult to put it back on.

8. Perhaps the most useful piece of advice for the novice builder is to ask for advice. The elderly, experienced salesperson at the local hardware store may be a fund of building knowledge. Don't hesitate to tap this valuable source.

The Site

The site that you have chosen for the greenhouse may demand attention before you can begin foundation work. In certain cases, a site that is not level can work to your advantage. If the terrain slopes away from the existing structure, for instance, you might consider "sinking" or terracing the floor level of the greenhouse interior (Fig. 47). A good depth for attached greenhouses is the same depth as the home foundation. One of the main reasons for excavating is to lower the "profile" of the unit so that it fits beneath the existing eaves of the house. Sinking the greenhouse will require more excavation than simply leveling the site, but it can result in more usable vertical space. Situating the highest point of the greenhouse interior directly adjacent to a house window or doorway will also supply more usable heat to the home.

If the ground slopes laterally to the side of the house, you may wish to design a split-level floor plan rather than level the entire site. Whether split-level, sunken, or used in its existing state, the site should be relatively level (side to side, front to back) before beginning foundation work.

The excavation depth for a greenhouse is determined by several factors. Many people have the misconception that if you dig a little way into the earth, the below-grade soil will act as thermal storage. Actually, you'd have to go down a considerable distance below the frost line to reach earth that would constitute a heat gain for a winter greenhouse.

"Pit greenhouses" or "grow holes" are based on this principle. They are dug out several feet below the frost line to enjoy the benefits of the earth's thermal storage. We have observed that grow holes perform slightly better than solar greenhouses *only* in extremely cold weather (below -20°F in the New Mexico region).

To achieve increased performance in a pit type of greenhouse adjacent to the home, you might have to dig down several feet below the foundation of the dwelling. This is *not* advised. If you have an existing deep cellar or basement with strong walls and good drainage away from it, an attached grow hole might do quite well. The hot air in the apex of the greenhouse would enter low in the home and the cool air in the basement would be circulated into the lower part of the greenhouse. The problem is that any design of this nature would require extensive excavation, landscaping and a thorough knowledge of the strength and condition of existing walls. It is not a recommended project for novice builders.

The Foundation

The foundation of any structure is one of its most important elements. If it is built properly, many future problems will be avoided. Careful measurements for the foundation are essential. We will assume for the purpose of these construction steps that you are building a greenhouse with a rectangular floor plan and that you are attaching it to the home. For the materials needed, see the list at the end of the chapter.

To determine the ninety degree corners off of the structure, place one edge of your framing square against the existing wall and extend the other edge with a string (see Fig. 48). Stake this string at the distance you have determined for the outer boundary of the greenhouse. After repeating this procedure for the other end wall, measure to see that the two strings are of equal length ("A" to "B" equals "C" to "D"). Connecting the outer perimeter stakes should produce a rectangle ("A" to "C" equals "B" to "D"). To double check your ninety degree corner angles, see that the diagonal measurements are equal ("A" to "D" equals "B" to "C").

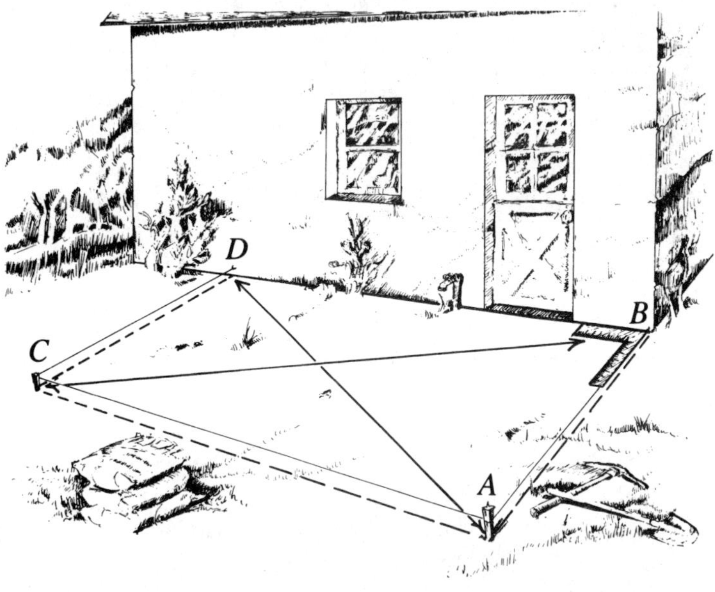

Figure 48

We will describe the poured concrete/rock type of foundation because it is widely used and easily understood and constructed by the home builder. It is appropriate in areas that have relatively shallow frost lines (three feet or less), and it will meet code anywhere.

Along the perimeter of the greenhouse excavate a trench to the desired width and depth. Make it at least 4 inches wider than the walls of the greenhouse. Massive stone or concrete/block walls, for instance, will require a wider foundation than frame walls. Drive stakes into the trench at 6 to 8 foot intervals, leaving 6 inches of the stakes exposed above the bottom of the trench. Check that the trench is level by laying a flat board from the top of the highest stake and taking a reading with the level from there. Do this around the perimeter of the trench. Fill when necessary; then smooth out the sides and bottom with a flat-nosed shovel. After the trench is dug, leveled and cleaned out, keep all interested gawkers away from the edges so they don't cave in the sides.

Before the foundation is poured, the outside of the trench should be insulated with 1" or more of rigid styrofoam (see Fig. 49). Cut the panels to size and fit them into the trench. They can be temporarily propped in place until the concrete is poured. Another method of insulating the perimeter is to wait until the foundation has been poured and the concrete has hardened; then dig a trench around the outside of it. Line the trench with sheet plastic and fill it with sawdust, dry pumice or styrofoam beads. Enclose the loose insulating material with plastic to keep it waterproof and cover the trench with dirt (Fig. 49).

Another prepouring step is to insert reinforcing material in the foundation trench. If you have to meet stringent building code requirements, this may be mandatory. "Re-bar" or "re-rod," as it's called, can be used in ½" or ⅜" diameter. It can be bought and cut to length at any building supply store. Two lengths of re-bar are laid along the bottom of the trench about 8 to 10 inches apart, supported 4 to 5 inches off the bottom by rocks. Ends that meet are lashed together with baling wire.

I've poured foundations with and without re-bar. I often throw as many river rocks as I can find (5 to 8 inches in diameter) into the bottom of the trench and forget about the re-bar (Fig. 50). I haven't noticed any settling or cracking in the foundations I've built this way.

The re-bar or no re-bar question reminds me of a typical bureaucratic hassle over the recent building of adobe homes in New Mexico pueblos. When the government engineers finally approved adobe for Indian housing (the all-adobe Taos pueblo has only been standing for a millennium or so), they stipulated

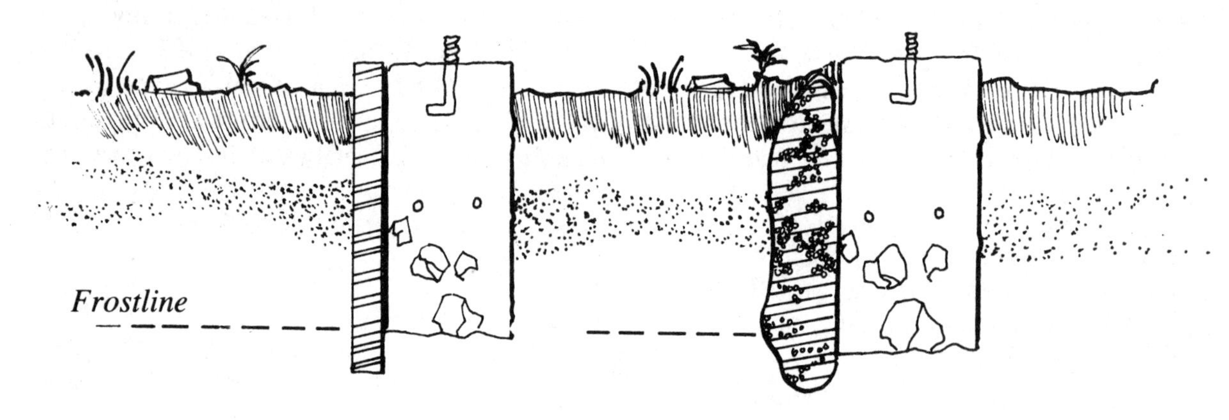

Rigid and Loose Foundation Insulation

Figure 49

The Old Carpenter's Water Trick

So, you want to check one end of the foundation trench with the other. You don't have a transit and the 2x4 won't bend. Get a friend. Then take a regular garden hose and lay it in the trench. Drive stakes in the end corners; each must be exactly the same height from the bottom of the trench. Holding the ends of the hose flush with the top of the stakes, fill it with water. If the extremes are level, the water at each end will be equal.

The beauty of this trick is that it will work for any length and over rough terrain (with a couple of people and plenty of garden hose). Of course, the hose ends have to be held higher than any point in between.

that re-bar be inserted in the vertical walls every several feet. A Santa Clara Pueblo friend of mine said, ''Isn't that going to frustrate the archaeologists a thousand years from now? They'll wonder what all the little red holes are doing in the middle of those mud walls.'' It's probably true that the adobe walls will be standing when the steel re-bar has rusted out. Suit yourself about the use of reinforcing bar.

It's a good idea to bring the foundation above grade (3 or 4 inches). This automatically eliminates some drainage problems and is definitely necessary if frame, adobe, or other water-soluble materials are going to be used to build the walls.

Old lumber can be used for the forms to restrain any concrete that is above grade. Most anything will do to secure them in place; large rocks, blocks, stakes, wire. Be sure the forms are the right distance apart and *well braced* so they don't spread with the weight of the concrete. Anyone who has worked with concrete can testify to its weight. Once you've had the terrifying experience of seeing a large mass of wet concrete start moving toward you, you'll always over-brace forms and wire the braces together.

Figure 50

On the inside of the forms, mark a level line for the top of the foundation. A chalk line works well for this. Make the line about 3'' above the actual level to which you are going to build the foundation so that it doesn't smear when the wet concrete is being poured. This is a convenient guide for a level pour (Fig. 51). Another way is to chalk line the exact level and height, and drive nails halfway in along the line. This gives an accurate guide.

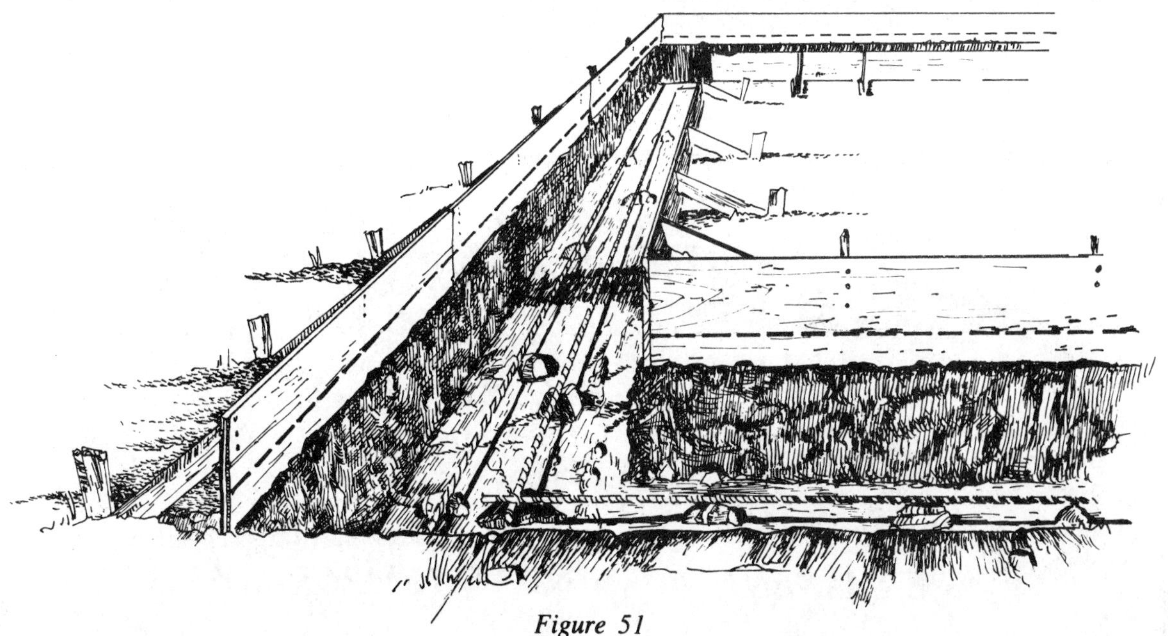

Figure 51

The fast way to pour the foundation is to have the ready-mix concrete truck back up to the site and dump it on you. But often the concrete companies won't deliver in small quantities, or the site is impossible to reach. In that case you have the option of buying premixed dry bags or making your own mix from cement and sand. If mixing your own (much cheaper), a standard concrete recipe is five parts sand and 3/4'' gravel (mixed equally) to two parts dry Portland, and water.

This is a heavy job so line up a few friends. The entire foundation should be poured at one time. You don't want to have seams from two or more separate pourings.

So, you're ready to do it. Sand and gravel in place, Portland bags stacked, shovels in the ready position, wheelbarrow greased, beer iced down. Consistency is what you want in the mix. It should not have dry clumps or an overabundance of any ingredients. The mix should be wet without being runny. If you pull a hoe through it, it should make nasty noises. When the mix is just right, it reacts like Jello when patted with a trowel. Nice stuff.

Start at one end of the trench and work around. After a load, usually a full wheelbarrow, is dumped, spread the concrete along the trench. Push and work the mix down into the trench with a trowel. Don't be gentle. You want to avoid holes or pockets in the foundation. Keep adding loads of concrete until you've nearly reached the level line or marker established as the top of the foundation.

As you work around the trench, pat and smooth out the top of finished areas. After the cement begins to set up, insert an anchor bolt (screw threads exposed) for sections such as low door jambs that will be framed above the foundation. With a square make sure that these bolts are perpendicular and in line with where the plate will be and that you've left about 1½ inches extending above the poured foundation (see Fig. 52). The plate, which we will talk about in various contexts throughout this chapter, is not something on which dinner is served. It is a piece of lumber, in this case a 2x4. The foundation plate provides a base for the frame walls of your greenhouse. The top plates give vertical studs and roof rafters something to hang on to. In general plates serve as weight supporting members of any frame structure.

Note: clean your tools immediately after use or they'll never be the same. If for some reason you have to leave a load in the wheelbarrow or mixer for a short time, pour a small amount of water on top of it and cover as tightly as possible. This also applies to mortar and plaster mixes.

After the pouring is done, check to see if any areas have sunk and make sure that the above-grade forms are secure. When the concrete has set up or hardened (usually within three or four hours), spray it

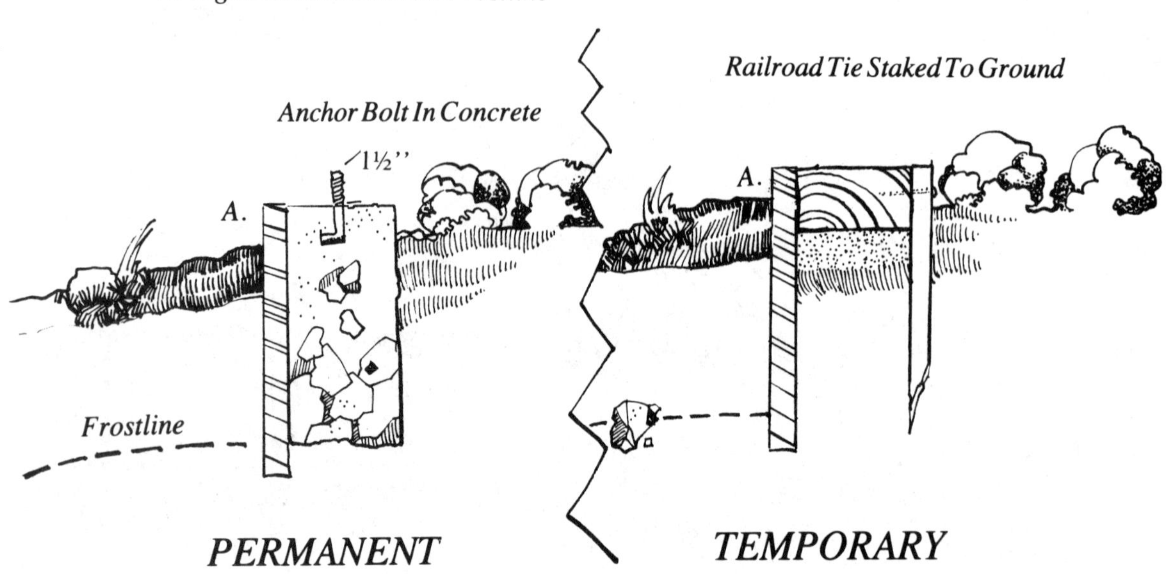

Figure 52

Figure 53

with a light mist of water or cover with wet hay or straw. This prohibits rapid evaporation that might crack or weaken the foundation. Spray it every few hours for the next day or two (don't bother at night).

When the foundation has a feeling of permanence, the forms can be removed. Clean well and recycle them into shelves, tables or bed frames for the greenhouse interior.

There are ways to avoid laying a foundation under the frame portions. I'll give one to you as an option. This method is common in large commercial greenhouse construction, and might be useful if you rent your home and plan a portable structure. Level the ground where the frame walls will be. (Don't dig a trench; just level the earth.) Wood plates (I've used railroad ties) are laid directly on two inches of sand and staked in at 3 to 4 foot intervals (Fig. 52). The stakes can be metal or wood but should be at least 36'' long. They can be screwed, nailed or bolted to the ground plate. The wood should be treated with copper naphthenate as a preservative (Fig. 53). Don't use *fresh creosote* or *pentachlorophenol* (''penta''), as these chemicals give off fumes that are noxious to plants.

Whatever foundation method you choose, the most important considerations are:

1) Is the weight evenly distributed?
2) Is the foundation level?
3) Will the water drain away from it?

If these criteria are met, the foundation will be functional.

Another commonly used foundation is the stem wall (Fig. 54), required by building codes in many areas. In regions having deep frost lines, the wide footing and narrow stem design can be more economical than a deep foundation of uniform width. The upright stem must be sized to receive the type of wall to be built above it (narrow for frame, wider for massive walls).

A related alternative to the stem wall is the pressure-treated wooden foundation pictured overleaf (Fig. 55). Here, treated 2 x 4s, 2 x 6s, and plywood panels take the place of concrete. The space between the vertical 2 x 4s is insulated and has a vapor barrier on the inside.

A third foundation option is the pier type of construction shown in Fig. 56. Individual piers of treated wood (or concrete) support a plate on which the greenhouse frame walls will be built. The pier foundation is perhaps the most economical one for areas having deep frostline.

In all of these foundation options panels of rigid insulation surround the perimeter to reduce heat loss. Construction steps and details for these foundations can be found in several of the books listed in the Bibliography.

An experimental option that could have wide application is presently being evaluated by solar architects at the Massachusetts Institute of Technology. In order to avoid the extensive trenching (and cost) required in deep frostline areas, rigid insulation is placed vertically around a relatively shallow foundation *and horizontally* away from its edge (Fig. 57). No frost heaving was observed during the first heating season (1977-78).

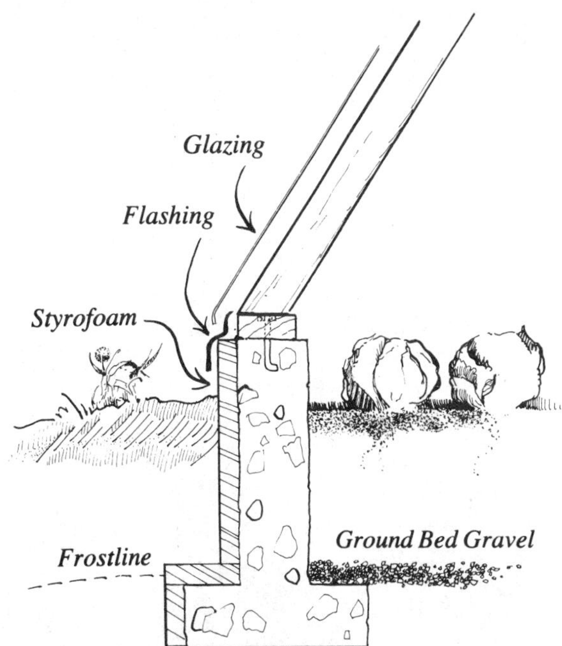

Figure 54

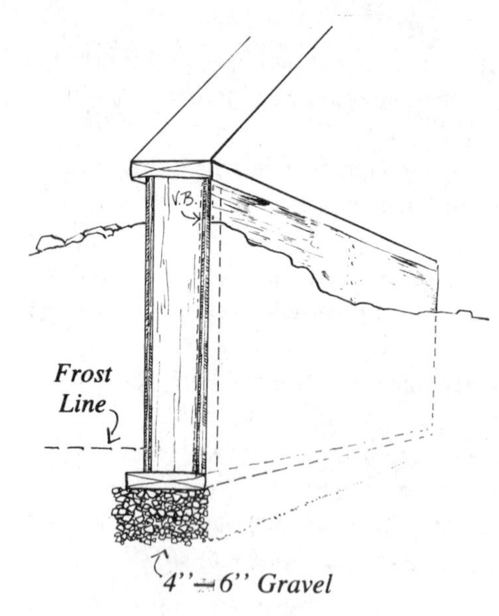

Figure 55

Affectionately known as the ''old New Mexico'' style, our final foundation alternative (Fig. 58) minimizes technological advances and expense. A bed of river rocks packed in a shallow trench of local mud is the weight-distribution medium. Larger, flatter rocks should line the trench bottom. A mud cap on top forms the flat surface for laying adobe bricks. (*Note:* This foundation can only be used in dry regions with deep ground-water levels and on sites with good drainage. It will pass *no* codes—nowhere—at no time, and is inappropriate for areas experiencing earth tremors or quakes. In many areas of the Southwest adobe homes have been standing on these foundations for hundreds of years.)

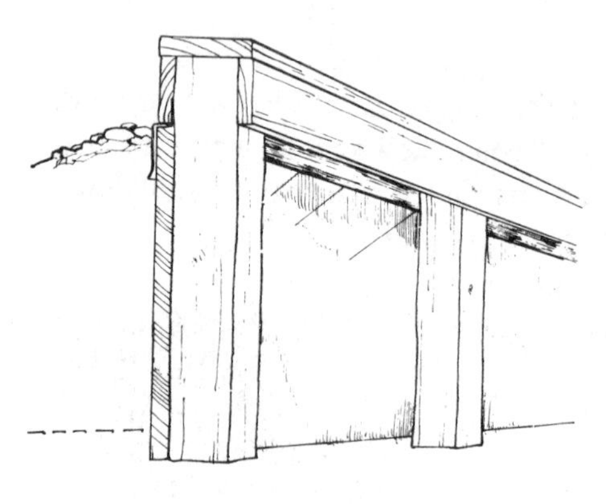

Figure 56

Massive Walls

When the concrete in the foundation has cured, you can begin work on the massive masonry walls. Different types of masonry constuction call for different techniques. Let's use hollow concrete blocks for our example. (Basically, the same technique applies to building brick or adobe walls, except that you can use mud for mortar in the latter case.) The standard size pumice block is 15-1/2'' long x 7-1/2'' high x 7-1/2'' wide. (They also come in half blocks and about every other size and shape imaginable.) For estimating the amount needed, use the dimensions 16''x 8'' x 8'' because of the added space to be filled by mortar.Before making your estimation, determine the exact size and location of any vents or doors in the walls. They must have a jamb or frame built around them, and that lumber is usually one and a half inches wide. Include twice that width (both sides) in your calculations.

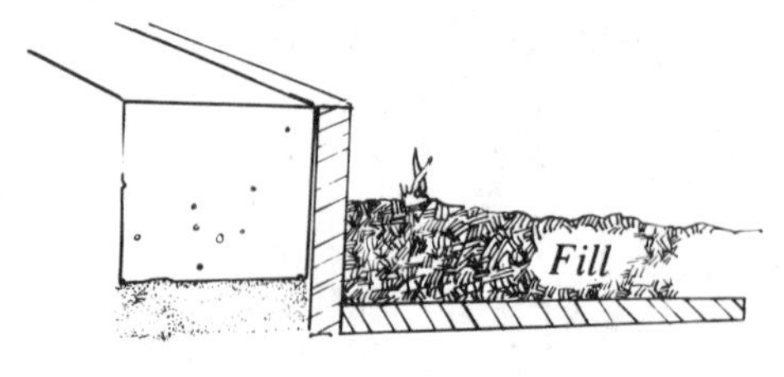

Figure 57

I like to avoid openings in masonry walls whenever possible. They involve a precision and degree of patience that I often lack. It's usually easier to locate vents and doors in areas that will be frame.

As the diagram in Fig. 59 (following page) shows, the high vent is set in the eastern frame wall (''A''). The low southwest vent sits on the masonry wall (''B''). The lowest part of the door is set in the east masonry wall (''C''), the upper 4/5ths in a frame section.

When the size of openings in masonry walls is determined, the estimate of the total number of blocks needed can be made. Determine the square footage of the walls and estimate one block per square foot plus 10 percent for cutting. For our example I'd buy 80 full blocks and 25 half blocks.

Figure 58

However, if you're having problems making the necessary calculations, remember that elderly, experienced salesperson we mentioned earlier. Write the dimensions of the walls (with doors and vents figured in) on a piece of paper and go to your local hardware store. Odds are you will get all the help you need and some good advice on the side.

You may want to order more than we've estimated. Masonry blocks look rather formidable, but are actually quite fragile and easily broken until they're in the wall. Handle and stack them carefully. If you do have leftovers, most reputable building supply stores will refund money on items returned in good condition and with a receipt.

The mortar used in laying blocks is made with masonry cement and screened sand. The standard proportions are 5½ parts screened sand: 2 parts masonry: 1 part Portland. Mixing equipment is the same as for the foundation. Small amounts are made as needed. A large triangular trowel is used to spread the mortar.

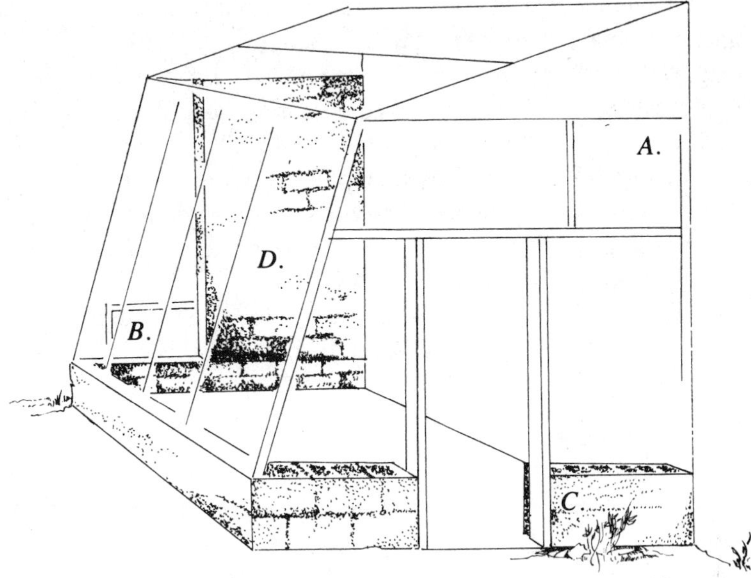

Figure 59

To lay a perfectly straight and level block wall, poles can be erected precisely at the outside edge of each corner of the structure. Marks are made up the poles at 8'' vertical intervals. Poles and marks are checked against each other with heavy twine and a string level. They should all be at exactly the same level for each course of blocks, and they indicate the top of each layer. That's the precise way to do it.

Another method is to simply begin by laying the blocks, checking for level, plumb, and straightness as you go. In an area as small as our 10' x 16' example, this should be sufficient. Lay the first course of blocks all the way around. Check for square on the corners. Now, using the level, build up two courses of blocks at each corner. String is strung between corners and straight runs laid down to the string. Checks for vertical can also be made with a carpenter's level or plumb line. If you're new at this kind of work, don't trust your eye too much.

Have you ever watched an experienced mason building a wall and seen the beautiful fluid movements he makes? A quick scoop of the trowel places a glob of mortar on the edge. The mason flicks his wrist downward, a microscopic spasm, and the mortar adheres to the blade in a flat, compressed mass. A long, thin line of cement is spread along a four-foot edge of the existing wall in the next stroke. Another similar motion, backhanded, lays it on the outer edge of the wall. The third trowel of mortar is applied in fast, choppy slaps center ridges. The new block is picked up and hit with mortar on two vertical edges, and wham, in it goes. This takes about six seconds. Like a finely tuned human machine, the mason progresses along the wall with the speed, economy of movement and accuracy found in downhill racers and basketball centers. The only things that slow down a pro like this are scaffold movers and unions.

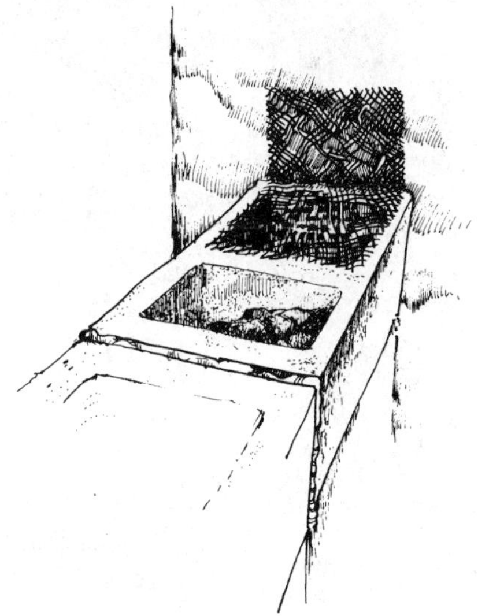

Figure 60

Don't expect to match this degree of skill. Try to put a uniform thickness of mortar (3/8 to 1/2 inch) on all seams. Any way you can get it to stick to the blocks is cricket. Try putting a small amount on the trowel and scraping it off with downward and outward motion. If you can't get the knack of this, put a larger amount on the trowel and shake it over the edges. I resort to my hands occasionally. ''Seat'' (firmly tap down) the block with the handle of the trowel. It should be evenly supported by mortar and level when it's in place. See that the seams are staggered, not one directly over another.

A method used to "tie-in" a new masonry wall to an existing structure is to bend a small (8 x 10") piece of metal lath to form a 90 degree angle and fit this between blocks in the new wall and the home wall (Fig. 60). Tack the lath securely to both walls. Lay the mortar and firmly seat the next course on top of the lath, pushing the block tightly to the existing wall. This should be done three or four times in a wall as tall as the one in our example, the 8-foot west wall.

Every three or four courses of cinder blocks, fill in the center holes with concrete (the same mixture you used for pouring the foundation). This strengthens the walls and adds mass, heat storage capability, to the structure. Before the concrete dries, insert anchor bolts where the framing plates will be applied to the top and sides of the walls (see Fig. 59).

When all the masonry walls are up, relatively straight, level and plumb, it's a good time to have a celebration. You can finally see and feel the results of your brain and muscle work. The hardest part is over. Enjoy it!

Frame Walls (Clear and Opaque)

Figure 61

As stated earlier, a frame wall does not have any appreciable mass; therefore it cannot store heat for a structure. However, when properly insulated, it will keep the heat in and it is often justified if other heat storage is planned. In the diagram in Figure 65, page 62, we show frame walls for purposes of illustration.

The vertical framing members (mullions) in the *clear-wall* sections are erected at 47-inch intervals (measured in what's called centers. Centers are determined by measuring from *one* edge to the *next* edge on the *same* side). The fiberglass that we recommend comes in 48" widths. The 47" center allows for overlapping the panels. Studs in the *solid frame* wall are on 24" centers, as are the roof rafters. All corner studs and upper plates are double-width for added strength. A scale drawing of your greenhouse using this type of specifications will help you estimate how many 2x4s of various lengths will be needed for the rough framing. Add the total linear footage of doors, vents and horizontal blocks (firestops) to the estimate. Add another 20 percent. We've supplied a list for this 10 x 16 foot unit at the end of the chapter.

You'll save money by having as few leftover scraps as possible. This is accomplished by making all principal framing members slightly shorter than an even number. Hence, if the mullion in the south wall is 7' 10" high, it can be cut from an 8' piece. On the other hand, if it is 8' 1" high, a 10' 2x4 usually must be purchased, and you're left with a 23" scrap. At current lumber prices, that's sinful. In your scale diagram, *make* the lengths come out economically. Do this by slightly changing angles and dimensions in the drawing until it works. If you just can't make it come out right, then plan to use the scrap lumber for tables, shelves, boxes, bed frames or other things. A 23" scrap, for instance, could be used for a firestop in the sheathed (solid frame) wall.

It's important to get a proper dollar value for your lumber. As with most things in life, you do this by choosing it yourself. Service personnel in the lumber yard are usually happy to let you choose and load your order. Occasionally it will be "company policy" not to let the customer look through the stock. Don't do business with a company like that.

Hold up each stud and look down the entire length for straightness. It should not be warped, twisted or badly bent. A slight curve is to be expected; it is called the "crown." If you are holding the stud on its edge, you can easily see the crown (Fig. 62). The crown is not to be confused with a bow in the broadside of the lumber which, if it's slight, can be taken out in installation. In construction, the crowns should all face the same way on the walls and should always be arch-up on the roof.

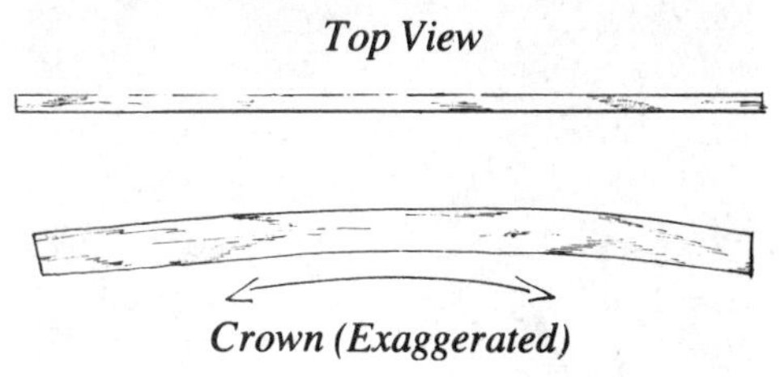

Figure 62

Next, check the stud for clearness. Does it have an unusual number of knots? Reject it. Does it have heavy "pitch" or "sap" areas? Reject it. Soundness can be tested by tapping it against a solid object. The stud should sound firm and full, not tinny or dead. When making all these tests, carefully place the rejects back on the pile or the yard people will never let you do it again. Lately, I get about a one-out-of-five acceptable ratio (that should give some idea of how much lumber you're moving around). Stud lumber, No. 2 common grade, won't be perfect, but get it as straight, clear and dry as you can.

After all this trouble in selecting the lumber, be sure to stack it flat and keep it dry when you get it home.

Figure 63

Let's begin by framing up the front face. First, you have the option of framing the wall in place or building it "prefab" on the ground (see Fig. 63 above). This decision is mostly a matter of personal preference. We're going to prefab the front face in our example, and frame the side walls in place.

Figure 64

In the diagram below, Fig. 65, the length of the roof and sill plates (top and bottom framing members) is exactly 16' (No. 1). The top plate should be double to prevent bowing under the weight of the roof. So the first step is to nail the two 2 x 4s together with 16 penny nails. Place the 2 x 4s flush and drive the nails in at an angle so that they don't protrude from the wood.

Lay the plates (bottom and top) on the ground and mark them where the vertical studs (47'' centers) will be attached. Ends should be double-width (No. 2); use 4x4s or sandwich two 2x4s together with ⅜'' plywood or wood lath in between to bring the dimensions to 4'' x 4'' (Fig. 64).

The next step is to cut the lumber to the appropriate length. In our example, we have seven mullions cut to 7' 7-1/2''. With the plates, this makes an 8' high front face.

Now if the south wall is to be vertical, the cuts are square on each end of the mullions. In our example, we have tilted the front face twenty degrees to make a seventy degree slope. (With any tilted south face, the bottom angle to the foundation plate will be the difference between the south face angle and ninety

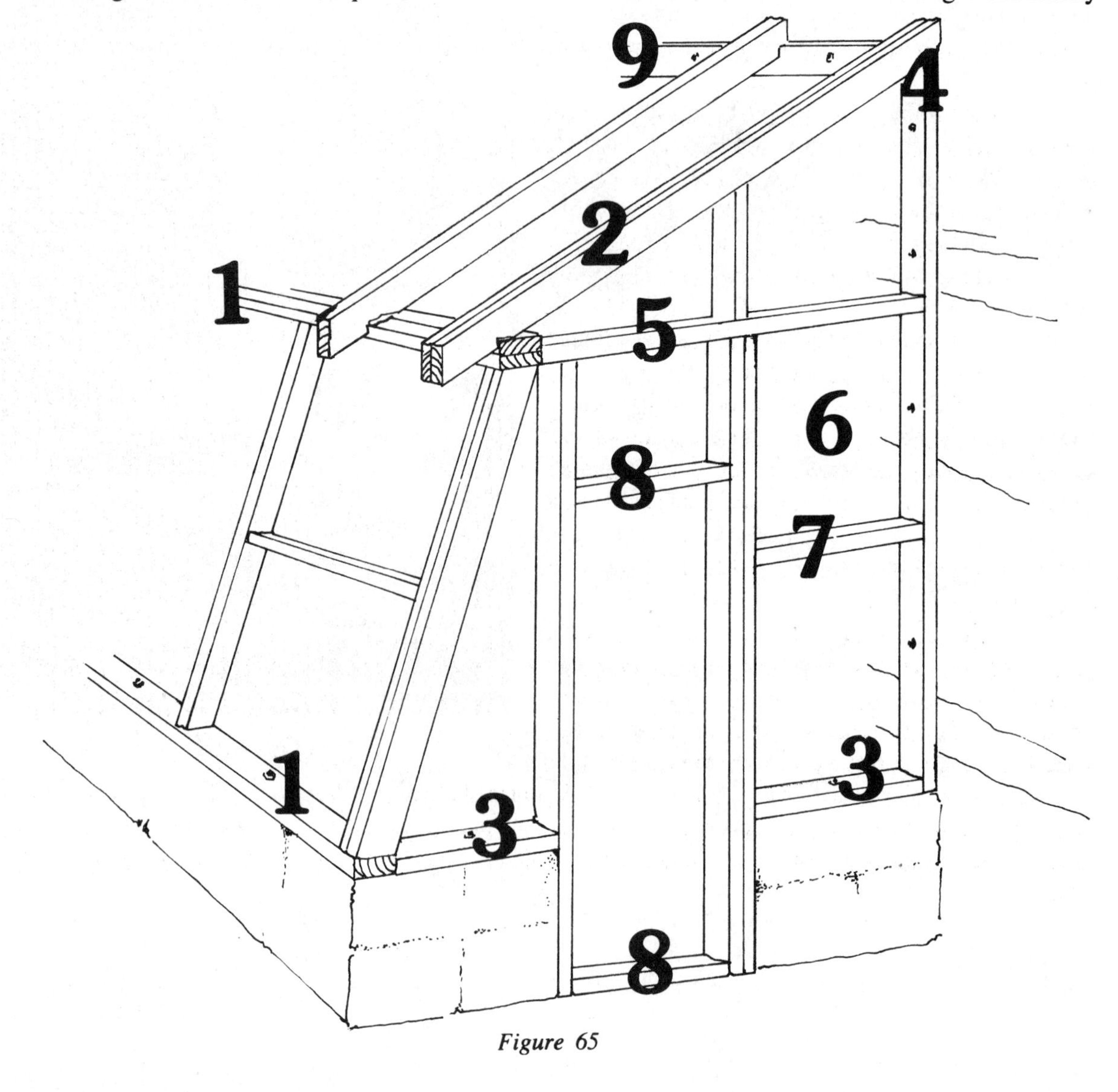

Figure 65

degrees.) All we need to do here is mark off twenty degrees from a ninety-degree cut. Make the top cut parallel to the bottom so that you have a horizontal surface for the top plate. When the mullions are cut, nail them at the marks to the top and bottom plates with 16-penny nails. Note: always buy high quality nails.

The south panel can now be lifted into place on the low front wall. This will take several folks working in unison. In order to precisely mark the junctions of the anchor bolts in the low wall with the bottom plate, gently lower the prefab frame structure into its permanent position. While several people hold the framework in place, one particularly aggressive type can pound on the plate with a hammer directly over the protruding anchor bolts to mark their position. Take the face off again and drill holes at the indicated marks. After installing metal flashing (see Fig. 54), the front face can then be installed on the wall, temporarily braced and bolted down.

At this point put up a few rafters (No. 2) for braces, the ones on the corners and a couple of others. With luck, you will be able to tie the roof rafters of the greenhouse into the existing rafters of your home. If not, attach a 2x6'' or 2x8'' ledger to the wall of the house as a base for the rafters (No. 9). Expansion bolts or large wood screws are used to get a *secure* tie-in to the wall. Don't scrimp here. The greenhouse roof must bear its own weight plus, in many areas of the country, snowloads *and* the additional loads of snow sliding off the roof of your house. Nail the rafters down to the top plate of the front face and to the plate or rafters on the house side (see next page for instructions). This will give the structure stability and give you a chance to see the outline of the greenhouse. When you have finished attaching a few rafters, take a break. Stand back and admire your work.

The next step is to cut and place the sill plates on the east and west walls (No. 3, Fig. 65). Bolt the plates down. To connect the frame walls of the greenhouse to the home, a ledger (No. 4) is securely "tied-in" to the home wall. It should fit snugly from the sill plate to directly under the rafter ledger plate (No. 9).

You now have the perimeter of the east and west walls to fill in with framing lumber. First install a plate that will also serve as a "header" for the lower frame walls (No. 5). In some cases the plate will also be the top of the door frame. Hold a piece of lumber level across the span and mark its intersection with the front face. On the west, it sits on the massive wall for the majority of the span. When the pieces are cut, nail them into place.

The east and west walls can now be framed in place. Remember, the clear sections will have 47'' centers. The insulated walls will be framed on 24'' centers (No. 6). The easiest way to do this is to mark the sill plate at the intervals where the upright studs are to be nailed. Take a carpenter's level and hold it against the side of the first stud (Fig. 66). Keep the base of the lumber on the bottom mark and get the stud exactly plumb. Then mark the point of intersection of the stud with the header on both the stud and the plate. Cut the stud about 1/16'' longer than the mark you've made.

Figure 66

All frame walls can be constructed in this way. It's easy and it's fast. Toenail (drive nails in at an angle) the studs as you go. When the studs are cut, fitted, and nailed, install horizontal firebreaks (No. 7), and the vent and door plates (No. 8). Make certain they're level, and nail them in with 16-penny nails.

Figure 67

Roof Rafters

The rafters for a short span (under 10 feet) can be 2x4's or 2x6's set on 24" centers. For longer spans use heavier lumber or put the rafters on closer centers.

Shallow notches are cut in the rafter ledger plate or joist hangers can be used instead. The rafters are toenailed to it at these points. At the intersection of the rafters and the south face top plate, very shallow notches (called "bird's mouth" notches) are cut in the bottom of the rafters so that they will rest snugly on the plate. Again, toenail them in (Fig. 67).

Putting up the rafters (and the roof) can force a person into some pretty strange acrobatic contortions. I would caution you that eight to nine feet up in the air is higher than you might imagine, especially from an aerial view. A body can be broken at the end of a free-fall from that height. Also, get in the habit of *not* leaving any tools or materials lying about on the rafters or the roof (even for a moment or two).

Figure 68

When all the rafters are in place, mark the intersection of the clear roof with the insulated roof areas. Use a chalk line to do this (see Fig. 78, p. 72). To determine that junction in your unit, use the charts as explained in Appendix B. It will usually be about halfway down the rafter span.

The figure on the following page illustrates another good method of joining the rafters to the front face top plate (Fig. 69). It can be used with sloped or vertical front-face configurations. In this design the roofing material will extend about three inches past the top plate. Depending on the roof length of your design, you may be able to economize on lumber by eliminating the rafter overhang recommended previously.

Another variation of the same join is pictured in Fig. 70. It features 2" x 6" rafters and front-face mullions that are cut to form a flush-fitting angle that is bolted together. The joined 2" x 6"'s are then trimmed to round off the sharp top angle (wood trimmed away is nailed to the inside of the join). This configuration will allow one continuous piece of flat fiberglas to serve as a clear roof and front-face cover, eliminating infiltration losses that can occur along the more conventional joins. The smooth contour it produces may also appeal to you and make this design aesthetically appropriate for your addition. It will be less angular in appearance than the other examples we have included. For details, see p. 109.

In contrast to this "knee joint" is the standard construction method shown below (Fig. 71). This type of rafter join is recommended for a vertical front-face configuration.

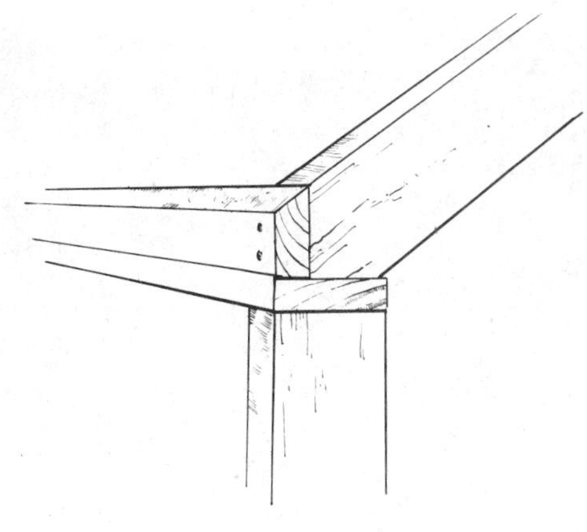

Figure 69

Painting the Frame

Rafters and framing lumber must be treated in humid climates. Copper naphthenate is recommended. Some paints have a preservative in them. After it's been treated, all framing lumber that will be visible within the clear walls and ceiling areas should be painted with a glossy white enamel or latex. This will help reflect light into the greenhouse and also enhance its appearance.

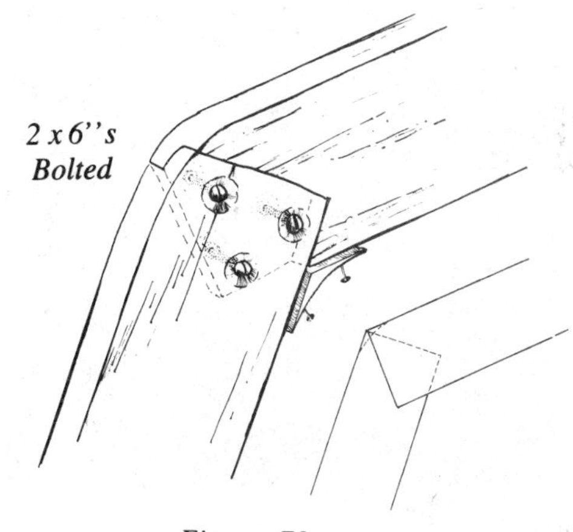

Figure 70

Choosing the Clear Glazing

Traditionally, glass has been used for the clear surfaces in greenhouses. Because of its resistance to high temperatures, it is also used extensively for covering solar collectors. Prefabricated sheets of double-layered glass are available from major manufacturers in various sizes as are many types of double glazed glass doors. The obvious advantages of glass are that it allows a view to the outside and is highly resistant to the harmful effects of weathering. It is, however, easily broken and demands extra care and considerable skill in installation.

Technological advances in the design and manufacture of plastics have produced important alternatives for the greenhouse builder. Some types of semi-rigid fiberglass/acrylic sheeting are guaranteed for 20 years to transmit enough light for photosynthesis. But, fiberglass can become cloudy or brown as a result of ultraviolet ray damage ("blooming"). Tedlar coating, however, has an ultraviolet retarding characteristic that helps greatly to preserve the clarity of the plastic to which it is applied.

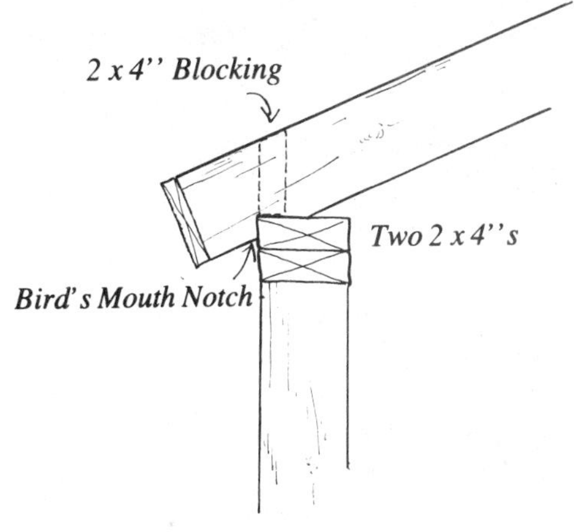

Figure 71

Fiberglass transmits nearly the same amount of light as glass, even though it's translucent rather than clear (see Fig. 72). Corrugated fiberglass is recommended for clear roof areas; it is easily installed and resistant to hail damage. Flat fiberglass is readily attached to clear-wall frames with rubber-gasketed nails or lath strips (see Fig. 73). I recommend the use of flat fiberglass on all vertical and near vertical surfaces. Flat material has a 20 percent smaller surface area than corrugated; therefore far less area for heat loss. The quality of light transmitted through fiberglass is

Figure 72

diffuse. It doesn't give the sharp, clearly defined shadow areas of glass. This is beneficial to plant growth. The cost per square foot of new fiberglass is considerably less than the cost of new glass.

There is an ecological question concerning the use of plastics in general. Plastic is a petrochemical product and is not biodegradable. The supply of petrochemicals is dwindling rapidly, and the atmosphere is becoming polluted by petrochemical wastes. Nondegradable products also constitute a form of pollution on earth.

I feel that our only hope for maintaining an ecological balance on the earth depends on a thoughtful, positive use of modern technology and its products. I believe that employing fiberglass and plastics in your greenhouse constitutes such a use. Many people are convinced that solar energy is the obvious fuel of the future. Further experimentation will lead to advances in the efficiency of solar collection, storage and transmission. It is probable that plastic products will become essential to the production of solar energy components. I suggest, therefore, that the intelligent and careful use of plastics in the greenhouse will serve as a valuable example of how to put our fossil fuels to beneficial use...before they are exhausted.

Material Properties of the Glazing

No glazing allows 100 percent of the total solar radiation through it. As mentioned previously, most glazings used in greenhouses transmit 80-90 percent of the light. The rest is either *absorbed* or *reflected* at the surface. The chart on pages 68 and 69 contains a wealth of information about common greenhouse glazing materials. The thickness of the materials is given in inches. *Solar transmittance* is expressed as a percent of the total light available at a normal angle of incidence. *Long wave transmittance* is a percentage of how much long wave energy is transmitted back out of the glazing once it gets inside. You'll notice most of the glazings are either at one extreme or the other. Glass is the lowest in long-wave transmission (it tends to hold in the heat), and polyethylene the worst at retaining energy. The *transparent* or *translucent* columns indicate whether you can see objects clearly through the material. Notice that some of the translucent materials are as high or higher in transmission than the transparent ones.

The visual impression you get looking through a translucent material is roughly equivalent to what you see when peeking through a wet shower stall door. The translucent glazings have the characteristic of diffusing the light passing through. They do this to varying degrees. The effect is that all of the light beam does not travel in a relatively straight line, as it does through clear glass. The beam scatters over a broad area, causing a more even light inside the greenhouse. Because the light is being scattered about the inside of the greenhouse, surfaces not in the direct path of the sun receive increased lighting. Several studies by agricultural research institutions have shown that plants respond better to this diffuse light regime than to direct light through clear glass. The scattered radiation will strike thermal mass in higher levels of the greenhouse and can be reflected more evenly by light-colored surfaces and bounced back to the plants below.

Maximum operating temperatures are the high-temperature limits a material can stand before it begins to deteriorate. A note here: these limits are usually set by the manufacturer who tests the materials under laboratory conditions.Glass may stand temperatures up to 400°F if it is mounted properly. Without adequate provision for its expansion and contraction it may crack if the collector reaches 250°F. Many people have been enamored with the transmission, low cost, and ease of application of greenhouse fiberglass. They will use it to cover a black flat plate collector that achieves 200°F temperatures, and then become upset when the glazing turns yellow in six months. All of the materials listed here can easily handle greenhouse temperatures. For high-temperature collectors or unusual conditions, please get information from the manufacturers. The weatherability, lifetime, and cost columns are self-explanatory. To update the cost for estimating, you might try adding on about 10 percent a year from the 1978 price listed.

The four most common materials used in home greenhouses are glass, fiberglass, acrylics, and polycarbonates. Because of their similarities in solar and long wave transmission, choosing one to use normally comes down to personal preferences in terms of initial cost, appearance, ease of application, and durability (as a function of cost).

We've worked with all of them and can give you some tips. All four glazings come in panels to cover large expanses. Maximum common widths are four feet.

Figure 73

Glass This is a beautiful, long-life material in the proper application. It's best in vertical walls in locations where vandalism won't be a problem. When you use it in tilted surfaces, stipulate safety or extra-strength glass. You can get good bargains in glass from local building contractors who have surplus patio doors. Glass can also be recycled from defunct greenhouses and urban renewal projects. Glass is not as difficult to mount as many people think, but *extra care* must be taken in terms of calculating exact framing dimensions and allowing for expansion. Don't be intimidated by glass; just respect its quirks.

Fiberglass We've used fiberglass more than any other glazing material because of its relatively low initial cost and its ease of application. Fiberglass is very forgiving stuff (it lets you cover minor mistakes in framing) and can be cut to any shape. See p. 70 for details of application. The drawbacks are lifespan and maintenance. After several years, the ultraviolet light in the solar spectrum starts to deteriorate the material. The small glass particles in the glazing become exposed on the outer surface and the material begins to look greenish-yellow. Most manufacturers sell a refinishing substance that is applied to restore the fiberglass to nearly its original clarity. We suggest you buy a brand with a twelve-year warranty and check the guarantee carefully.

Acrylics We've used the double-walled panels (4 feet wide x 8 feet, 10 feet, 12 feet long) and have been quite pleased with the results. The panels are excellent for roof sections where you want an attractive, light-diffusing surface. In high ultraviolet conditions, don't get the very thin panels; specify the thicker, more durable material. Allow for more expansion/contraction than glass (1/2'' in a 4' span). You can use the expensive mounting hardware sold by the manufacturers, or construct your own. Acrylic panels are very lightweight, and so, a joy to handle. One drawback is that they scratch quite easily. Be careful when moving them around. Follow the manufacturers recommendations to the letter in cutting and applying the material.

Polycarbonates These come in all thicknesses from films thinner than polyethylene to rigid panels used for bullet proof glass. As interior glazings anything from .005''-.060'' is recommended. For exterior, use .060'' or thicker. A polycarbonate can be used when the transparency of glass is desired but vandalism or breakage may occur. Although all polycarbonates begin as transparent as glass, weathering on exterior

A Comparison Of Glazing Materials

Courtesy New Mexico Solar Energy Association

Type	Comments	Brand Names	Thickness (in.)	Solar Trans-mittance (%)	Long Wave Trans-mittance (%)	Optical Clarity: Transparent	Optical Clarity: Translucent	Maximum Operat-ing Temp. (°F)	Weatherability	Lifetime (Years)	Cost / Sq.Ft.[2]
Glass	Pros: Excellent selective transmission. Transparent. Excellent weatherability. Resistant to U-V, air pollution & heat. Good insulation (Special glasses only). Low thermal expansion. Cons: Expensive. Breaks easily. Heavy. Installation relatively difficult.	2 Categories: -Window -Special glasses: 1.Double strength float 2.Thermopane	.125 .625	89	3	X		400°	Excellent.	Very long - 20+	.75-1.50 1.00-2.25 4.00-9.00
Acrylics	Pros: Excellent optical clarity--transparent available. Excellent weatherability. Good impact strength. Lightweight--easy to handle. Good insulation (twin wall only). Insulation & light transmission qualities equivalent to glass (Acrylite SDP). Cons: Expensive (but less than glass). Susceptible to surface abrasion. Heat sag at high temperatures (> 200° F).	Single Wall: -Plexiglas -Lucite -Acrylite Twin Wall: -Acrylite SDP -Cor-X-Acrylic -220 Twinwall -Polylux	.125 .625	~92 85-92	N/A N/A	X	 X	180-200 160°	Excellent. Excellent.	20 20	1.00-2.00 2.00-3.00
Polycarbonates	Pros: Very high impact resistance (used in jets, prisons) High service temperature. Like acrylics (but slightly reduced solar transm. Transparent available. Cons: Relatively expensive. Scratch easily. Non-rigid.	Merlon Lexan Tuffak Twinwall	.125 .22	86 82-89	6 N/A	X	 X	250-270	Slight color change & slight embrittlement with prolonged sunshine--for all polycarbonates.	15-17 5-7	2.25-3.25 1.25-1.75
Fiber-Reinforced Polyester (Fiberglass)	Pros: Low cost. Good solar transmission. High strength & durability (especially with protective coating like Tedlar). Easy to handle. Even diffusion of light. Cons: Transparent not available. Medium-range lifetime. Tedlar-coated has 8% less solar transmission.	Lascolite Filon Sunlite Flat: -Standard grade (not u-v resistant) -Superior grade (u-v resistant) -Supreme grade (Tedlar coated) Corrugated: -Standard -Superior -Supreme	.040 .040	79-89 85-90	6 10		X X	160° 200[3]	Slight yellowing with prolonged exposure. Also "blossoming" of glass fibers--for all fiberglass. Without u-v absorbers &/or protective coatings (standard grade), sunlight effect increases greatly.	4-7 7-13 13-20 Same as flat.	.25-.50 .40-.55 .55-.80 .25-.50 N/A .45-.60

Polyethylene	Pros: Very low cost. Light, flexible--easy to install. Good inner glazing.	Regular (not u-v resistant)--at any building supply store.	.004	>85	70-80		X	140°	Poor. Improved by: -using as inner glazing rather than outer -using u-v resistant brands.	4 mos.-1 yr.	.01-.03
	Cons: Short lifespan. High transparency to long-wave radiation. Wind & temperature sagging effects.	U-V Resistant: Monsanto 602								1-3 yrs.	.06-.10
Polyesters	Pros: Low cost. Superior surface hardness.	Mylar	.003-.014	85	16-32		X	300°	Discolors slightly with prolonged exposure.	1-5	.09-.35
	Cons: U-V degradable unless coated. Fairly high long-wave transmittance.										
Polyvinyl-Flourides	Pros: Excellent weatherability. Strong. High solar transmittance. Often bonded to fiberglass as U-V and weather-resistant screen.	Tedlar	.004	90-94	43		X	325°	Excellent.	~25	~.80
	Cons: Fairly high long-wave transmittance. Expensive. Only available up to 4 mil thickness (film).										
Fluorocarbons	Pros: Exceptional resistance to heat, weather, & chemicals. Very high solar transmittance.	Teflon FEP	.001	98	N/A		X	400°	Excellent.	25	~.59
	Cons: Expensive. Available only in films.										
Others: Vinyls	Clear films, more expensive and longer lasting than polyethylenes, but problems with U-V and dust.										
Cellulosics	Tough, with moderate thermal stability; U-V degradable unless coated; usually not easily available.										
Others	Plastic-plastic & plastic-glass laminates; double-walled channeled plastics, expanded plastic honeycomb plastics, clear insulative plastic films with air bubbles.										

Sources: Knudtsen, Peter K.; "Glazing Materials: Some Alternatives"; Southwest Bulletin, February, 1978. NMSEA Workshop Crew Handbook, chapter on glazings.

Note: For each of the 8 major glazing types, performance data is provided for at least one commercial product of that type. The other products of that same type will vary slightly among each other, but will basically perform similarly to the cited example.

Footnotes:
1 Most glazings are available in a variety of thicknesses. These figures indicate the thickness described by the data in the following columns.
2 Cost figures obtained in August, 1978, from retail distributors in New Mexico.
3 Manufacturer's claim only. Lascolite estimate cross-checked with other, non-manufacturer estimates. Both estimates refer to superior grade (u-v resistant) fiberglass.

PERFORMANCE DATA HERE IS LARGELY MANUFACTURER'S DATA -- Actual performance may be somewhat less.

Table IV

surfaces reduces their transparency with time. Polycarbonates have a tendency to whiten a bit on the outer surface, become slightly brittle and scratch easily. We prefer them as inner glazings behind glass and as storm windows. Many people have single-glazed glass greenhouses and want to keep a view. A thin-film polycarbonate makes an excellent inner glazing that can be removed and stored at the end of the winter season. The film, when applied as an inner layer or storm window, has the visual quality of old glass (slightly wavy) which we find quite lovely. The film comes in 2 and 4 foot widths and can be cut to any length. Panels are generally sold in 2-foot wide increments.

There is no *one* correct glazing. Many people are combining materials in home greenhouses. For instance, all four materials mentioned above plus polyethylene can be used in this manner: major exterior panels are fiberglass (for cost and light-diffusing quality); interior glazing is polyethylene (cost); the roof is acrylic twin wall (light diffusion, strength); exterior transparent windows for views are polycarbonate film (cost, ease of installation). A combination we've also used is thermopane glass for large vertical panels in the greenhouse and acrylic twin wall panels or double-glazed fiberglass for the clear roof sections. This gives the beauty of glass combined with the light-diffusing and structural qualities of acrylic or fiberglass. Many manufacturers will send samples upon request. Get to know a material before you stock up.

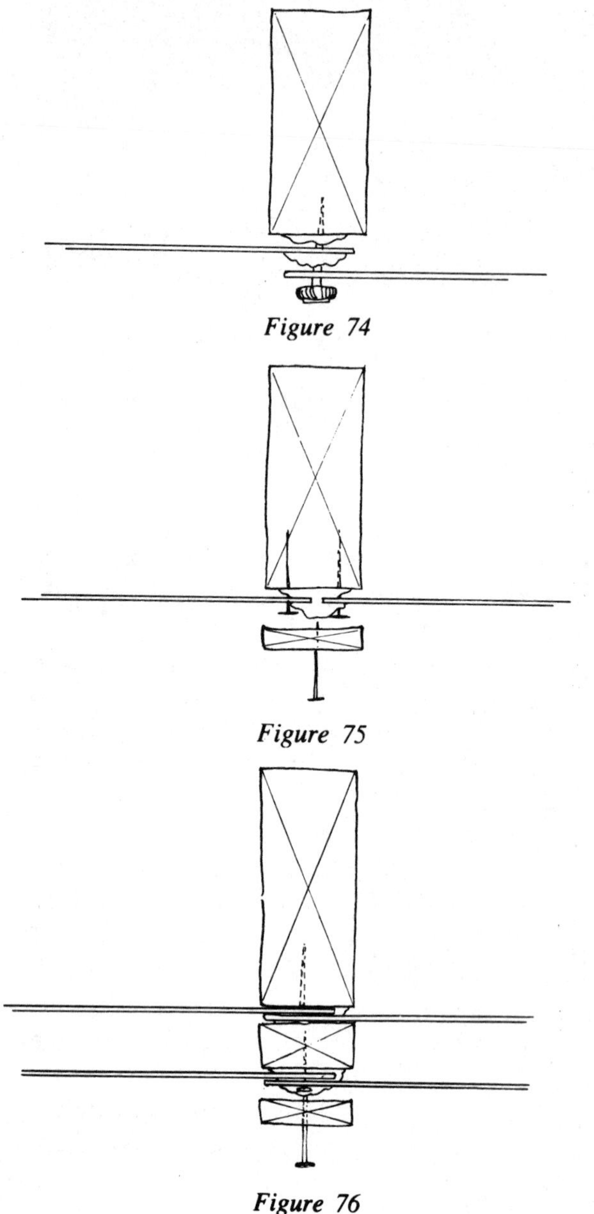

Figure 74

Figure 75

Figure 76

Clear Walls

Having built and painted the frame walls, you are now ready to install the outer layer of fiberglass. Most of the products (Lascolite and Filon, for example, brands we've used and can recommend highly) are sold in 4-foot-wide rolls, up to 50 feet long. For this reason, we suggest that you install one 4 foot panel at a time. Beginning at one end of the frame wall, measure the height of the section carefully. After measuring and marking the material with a felt-tipped pen or razor knife, cut the section with sharp wire-cutting shears. Apply an even bead of sealant, silicone or butyl rubber to the wood frame you are going to cover. With two or three people holding the edges of the material, align it in front of the frame. It is handy to make corresponding marks on the fiberglass and the wood frame to aid in alignment. Nail the top edge to the upper frame member in the center. Use a gasketed nail; do *not* nail it in completely. Check to see that the sides and bottom will fit properly. Nail the outside edge in the center (again, halfway in). Pulling slightly on the remaining two edges (bottom and inside), nail them temporarily in the centers of the frame. You can use flat-headed 1'' nails for the inside or leading edge or simply lay it on with no nails. The next section of plastic will overlay it and must fit snugly. The rubber-gasketed nail heads protrude about 1/8'' from the wood and would cause bumps in the overlaid plastic. Nail the material down to the horizontal braces first. Then, pulling diagonally on the four

Figure 77

corners of the sheet, put in nails approximately 8 inches apart and angle them out to pull tension on the sheet. Work the bulges out of the plastic from the center to the corners. If a major bulge has developed, try to detect it early; remove the temporary nails and realign the sheet.

After cutting the second panel, lay another bead of adhesive over the leading edge and the top and bottom. Proceed as above, overlapping the second panel by 1 inch on the edge previously flat-nailed. Use gasketed nails to attach the overlapping section (Fig. 74). Follow the steps given for the first panel, again using 1'' flat-headed nails in the leading edge to be overlaid by the next panel. For vents and removeable panels, install the fiberglass on the ground.

There are two very expensive items used in this method, silicone sealant and gasketed aluminum nails. Lately I've used a cheaper method that I believe is just as effective. Check and mount as before. Don't use any silicone sealant. Get small galvanized nails and drive them into the fiberglass about every six inches. After all panels are up, cover edges and overlaps with thin (¼'' x 1½'') wood lath (Fig. 75). If there are any bulges or leaks, they can be sealed with regular caulk.

You may choose to mount both glazings on the front of the south-facing clear wall (Fig. 76). This option will produce an effective interval between the layers of (rigid) fiberglass and allow for the installation of movable insulation within the space between the 2 x 4 mullions (see Fig. 26, p. 33).

The irregular shaped panels in the corners of the east and west wall are sized by cutting a rectangular piece to fit the length of the frame. Hold the piece up to the panel and mark the shape with a chalk line. Save the scraps for other irregular shapes and vents.

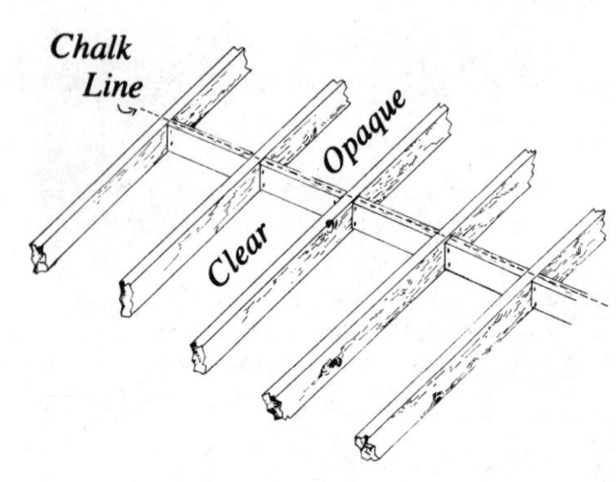

Figure 78

Roof Installation

Blocking (No. 7 in Fig. 65, p. 62) like that used between the vertical wall studs will be nailed between the roof rafters to receive the corrugated plastic and solid roofing material. Be sure to toenail-in a straight row of these blocks along the line separating the clear from the solid room (Fig. 78). In all other roof areas, the blocking can be staggered for ease of nailing (Fig. 79). Install 1/2 round molding along rafters and "wiggle board" across tops of roof-dividing and front face purlins (Fig. 80, below).

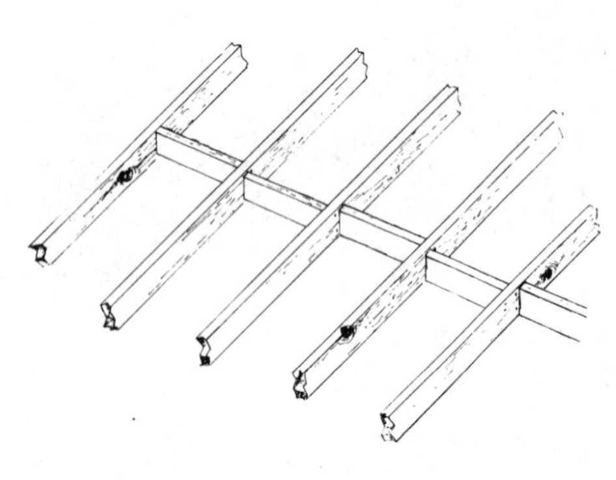

Figure 79

You are now ready to install the corrugated clear roofing. One sheet of corrugated is held in place at a corner. Align it carefully along the bottom and side edges. The roof can overhang the side two corrugations and the bottom about 6" unsupported. Get it "true," that is, perfectly in line, with the rafters and "wiggle boards." An out-of-line first piece will throw the roof whacky.

Now nail down the fiberglass every 3 to 4 corrugations with gasketed nails to the front face purlins and rafters. Nail into the high ridge rather than the valley. This will prevent water seepage through the nail hole. Overlap the sheets at least one and, if possible, two corrugations and continue across the clear area in this manner. Don't nail the top edge (to be overlapped by the opaque roof) yet. The only real trick to nailing into the fiberglass is to have a steady hand and a good eye. A missed hammer blow can splinter the material. leaving an ugly opaque mark (to say nothing of the damage to your thumb).

The solid roof is installed next. If plywood and composition roofing material is to be used, lay a strip of corrugated foam molding across the top edge of the clear/solid roof junction. Nail the plywood sheets onto the rafters, overlapping the foam strip and the clear roof area by 3 to 4 inches. The foam molding above the corrugated plastic will create a tight seal against heat loss. You can now apply your composition shingles or other roofing material over the plywood (Fig. 81).

Vents in the roof are *not* recommended for any but the most experienced builder. It is extremely difficult to seal them against air and water leaks. If you have the knowledge, build a "boat hatch" type vent of 1 x 4s that extends about 2" above the roof. Fit the vent so that it lays flush on the 1 x 4 box.

In much of the country a common roofing materal is corrugated galvanized steel. It is relatively cheap, easily installed and maintenance free. If you use metal, no plywood is needed. Simply align the corrugations of the galvanized with those of the plastic (they match) and nail it down to the rafters and blocking. Again, the solid roof should overlap the clear areas. For nailing, hold lead-headed roofing nails with a pair of pliers over the designated spot (on a ridge and over a rafter or block) and strike solidly with a hammer. Wear eye protection.

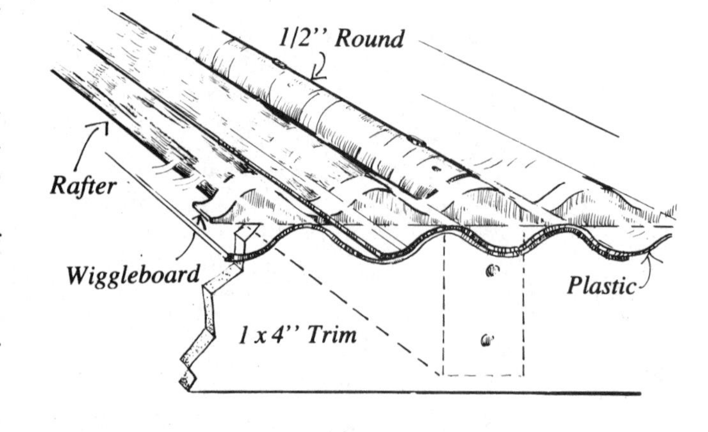

Figure 80

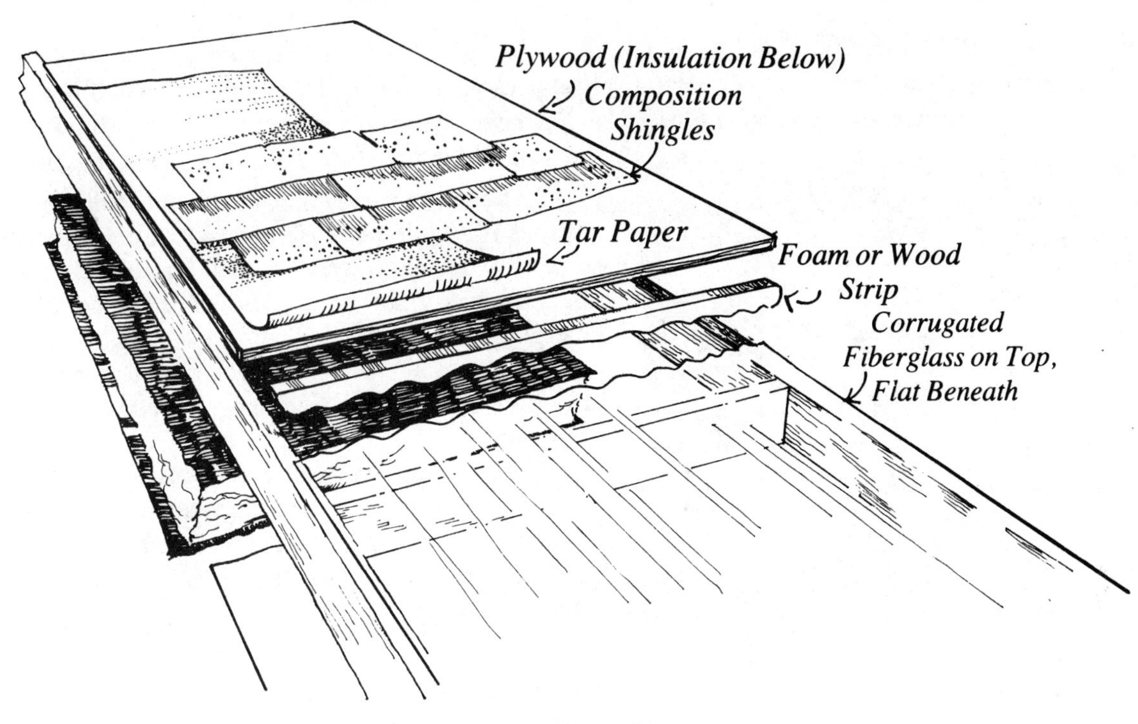

Figure 81

The Interior Glazing

Depending on the framing method you have chosen for your clear walls, the interior plastic will require one of the installations pictured at right (Fig. 82). The three framing options range from most economical (A) to most expensive and durable (C). Example B is the method recommended in the framing section.

Six-mil polyethylene (with an ultraviolet inhibitor) is suggested as an inexpensive glazing for the interior clear walls and roof areas.It is susceptible to weathering but will be protected by the fiberglass outside layer; it should last three to five years before having to be replaced. I've used both Monsanto 602 and Tedlar.

The thin plastic is more easily installed than the outside fiberglass. Large sections of clear area can be covered at once. Using scissors or a razor knife, cut out the sections to be attached. Make your cuts at least 3 inches longer than measured. With three or four helpers holding the extremities of the sheet, begin stapling along the centermost mullion. Work outward toward the sides and corners, stretching the sheet as staples are driven 8 to 10 inches apart. Wood lath (¼" x 1½") is nailed over the plastic-covered mullions with finishing nails to produce the final seal. You can stain or paint the stripping before nailing it up to make it more attractive.

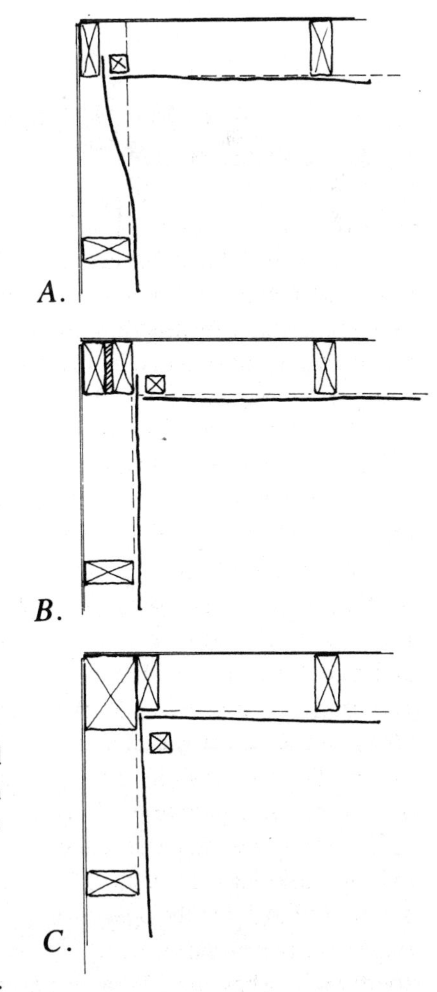

Figure 82

Insulating Frame Walls

Solid frame walls can be insulated with a wide variety of materials. Rockwool, fiberglass, styrofoam, polyurethane, pumice, and cork are all excellent. Foil-backed insulation stops a lot of radiant heat loss through the walls and acts as a vapor barrier, but it can be prohibitively expensive. If you do use it, install it with the foil backing facing the interior of the greenhouse. Recently I've been stapling heavy duty tinfoil on the interior side of the fiberglass batt. That's cheap and effective.

The amount of insulation applied should be as much as is used in the walls and roof of a well-built home in your area. I usually use at least 4 inches of fiberglass batt in the walls and 6 inches in the roof. Fiberglass is easily applied with a staple gun or tack hammer. Salespeople at the local hardware store can help you choose the right amount and type of insulation.

Always wear a long-sleeved shirt, gloves, and button your collar when installing fiberglass insulation. Wear safety glasses for rafter work, especially if you have sensitive eyes, and try not to breathe too much.

Figure 83

Interior Paneling

Insulated frame walls can be paneled with any material you find attractive. Masonite, plywood, waterproof sheetrock, and rough or finished lumber are widely used. Paneling materials call for various adhering techniques. Wood products are usually simply nailed with finishing nails. Some paneling can be glued with construction adhesives. When using large panels, measure carefully before cutting. If you're careful, you can save odd-shaped scraps for other areas and keep waste to a minimum. Occasionally you may have to nail in a "scab," a scrap of 2x4, to the studwalls in order to give a nailing surface for the paneling. If corners don't "flush out" perfectly, don't worry. That's what God made corner lath for.

As the insulated wall will not store heat, the inside should be a light color to produce a reflective surface. Water-sealing will help to protect the interior against deterioration due to high humidity.

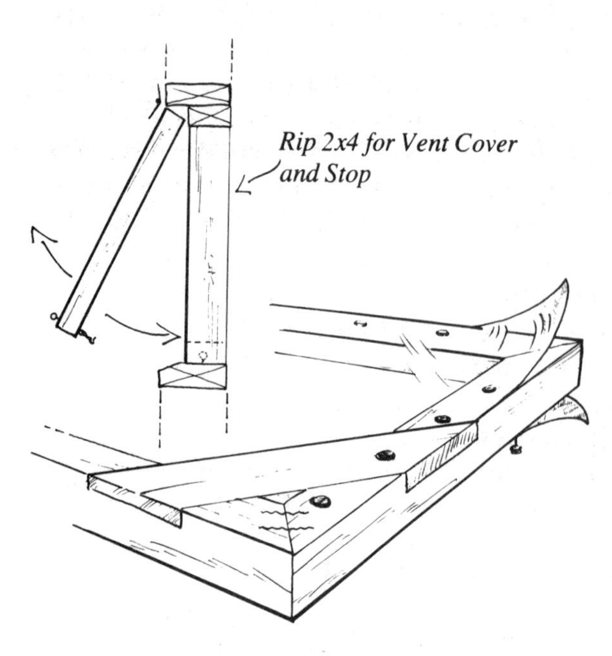

Figure 84

Doors and Vents

Doors and vents must be tight-fitting and weather-stripped around all joining surfaces. If you are using fiberglass for the large collecting surfaces, you may choose to use it for the vents and/or doors. Clear views to the outside can offer attractive accents to opaque and translucent walls. When glass is used, it should be double-glazed to reduce heat loss. If possible, windows and doors should slide or hinge away from the greenhouse interior. Hinges are mounted on top of vents to open out (Fig. 84). Remember, the high vent that you put in should be on the downwind side of the prevailing air flow.

In determining the dimensions for an outside door, remember that you will be moving large quantities of soil into the greenhouse. Make the door wide enough to accommodate a wheelbarrow (plus your knuckles). Thirty-two inches is good. It is also advisable to make the height of the door a standard measurement for convenient access. In very cold climates an air lock over the exterior door will save large heat losses (see Herb Shop, p. 140).

Exterior Paneling and Insulating

All massive walls should be insulated on the outside. An effective insulating material for this is styrofoam or styrene panel (1'' or 2''). They can be stuck to the walls with a heavy duty construction adhesive. Use it liberally. If the wall is to be plastered, cover the styrofoam with tar paper. Then use firring nails and chicken wire over that. The wall can now be plastered with a hard coat (5 parts sand: 3 parts Portland: 1 part lime).

Exterior covering of the frame walls can be any material that suits your aesthetic and economic criteria. I've used old lumber, plywood, Celotex (exterior fiber sheathing), and metal siding. The most important thing is to make sure there are no leaks that allow water into the frame walls. Remember—higher panels overlap lower ones for waterproofing. Caulk anything that looks suspicious.

Sealing the Greenhouse

If there is one most critical factor in the construction of your solar greenhouse, it is that *all joining surfaces fit tightly together*. Tightness to reduce air leaks is one of the highest goals which you can attain, both as a designer-builder and proud owner. But another important goal is to produce an attractive environment that reflects the care and thoughtful work that went into the design and construction.

During each phase of building, ask yourself where heat loss is likely to occur. Wherever large surfaces are joined (such as the foundation plate to the foundation), a layer of fiberglass or foam insulation will help solve the problem. Sealants such as silicone and caulking materials (wood filler, spackling compound) should be used on any crack or opening that could transmit air through the structure. For larger openings, such as might occur at the house/greenhouse junction, use metal lath and plaster to build airtight walls (insulate between them). As we mentioned earlier, vents and doors must be completely weathertight. The importance of sealing insulation cannot be overemphasized, as it can make the difference between the success or failure of your greenhouse.

THERE YOU ARE! YOU DID IT! CONGRATULATIONS!

Annotated Materials List

This is an outline of construction materials commonly used in Solar Sustenance Team/New Mexico Solar Energy Association solar greenhouse workshops.
Use it as a general checklist for similar projects whether they be greenhouses, workshops or home improvements.
The "Quantity" column is for a 10' x 16' floor size; it can be modified for other dimensions.

Quantity	ITEM	USE
	Foundation	
48 ft.^{3}	Rocks (2" to 6" diameter)	Fill most of footing trench (2' deep by 6-8" wide).
5 sheets	Styrene beadboard insulation (1-2" thick).	Line outer edge of trench. Reduce heat loss.
3 sacks	Portland cement	Mix with some rock; makes concrete footing 8" wide x
6 ft.^{3}	Concrete sand, #4 steel re-bar	6" deep at top of trench. Center re-bar.
10	Concrete anchor bolts	Place every 4'. Tie wood frame to block walls.
80	Full blocks	Perimeter block walls.
25	Half blocks	Perimeter block walls.
	Frame	
36'	2 x 4 sill plates	Bolt to concrete footing.
130'	2 x 4 studs	Skeleton of walls.
72'	2 x 4 rafters	Skeleton of roof.
48'	2 x 4 plates (wall, ledger, top)	Tie roof to front and to house wall.
110'	2 x 4 blocking (firebreaks)	Fill in frame grid.
	Finishes & Fasteners	
40'	6" galvanized flashing	From sill plate, over buried insulation
6 lb.	16 penny nails	basic framing
4 lb.	8 penny nails	siding, light framing
2 lb.	3 penny galvanized nails	fiberglass, batten, exterior trim
2 lb.	6 penny galvanized nails	exterior trim
1 lb.	dry wall nails (glued, ring-shank)	gypsum wall board
1 pkg.	1/4"-3/8" staples	vapor barrier, inner glazing
2 lb.	lead-head roof nails	steel roof only
150 count	rubber-washered aluminum nails	fiberglass roof only
5	rafter hangers with nails	rafter—wall plate
8-10	60 penny spikes or lag screws	wall plates—house wall
1 sack	wall board joint compound and tape	seal wallboard edges and texture surface
6 pair	3-1/2" square flush hinges	door hinging
3-5 tubes	butyl or latex caulk	sill—concrete, general sealing
2 tubes	butyl gutter caulk	metal roof seams
2 tubes	clear silicone caulk	seal fiberglass edges
2 gal.	white exterior latex paint	all wood exposed or under glazing
2 qt.	dark (black) paint	water storage barrels
40-60'	1 x 6 pine boards	exterior trim
1 tube	wood glue	door, vent frames
1 gal.	cuprinol or copper naphthenate	preserve wood

	ITEM	USE
	Roof	
4-8'	Corrugated steel sheet (2' wide)	Exterior surface
32'	Insulation batts (2' x 3-1/2'')	Under steel roof
70 ft.^{2}	Polyethylene 4-6 mil film	Vapor barrier under batting
2 sheets	Gypsum wall board (3/8'')	Inner ceiling
	Insulated Walls	
80 ft.^{2}	15 lb. building paper	Moisture-resistant barrier outside bolts
3	Exterior siding (4' x 8' sheets)	Exterior sheathing
40'	Insulation batts (2' x 3-1/2')	Fill wall cavities
80 ft.^{2}	Polyethylene 4-6 mil film	Vapor barrier inside batts
3	Gypsum wallboard (3/8'' x 1/2'')	Inner wall
	Clear Walls & Roof (South half of greenhouse)	
2-10'	Fiberglass-corrugated (4' wide)	Front half roof
1-50' roll	Fiberglass-flat 4' wide (4-5 oz/ft.^{2})	Outer wall surface
200'	Wooden lath (1/4'' x 1-1/2'')	Batten strips at seams
200-250 ft.^{2}	Clear Polyethylene, 6-8 mil ultraviolet inhibited film.	Inner wall surface
50'	corrugated closure strip	Set on rafters. Match roofing corrugations.
40'	1/2-round closure strip	Cover base of front wall. Hinged air in-lets.
1	Wood storm door, glazed 2 sides with fiberglass	Translucent door to outside
	"Furniture"	
6	55 gallon or 30 gallon drums	store heat in water (closed lids)
	scrap boards	2' deep planting boxes
	pea gravel or flagstone	floor outside beds, barrels

CHAPTER VII THE GREENHOUSE GARDEN

Solar greenhouse gardening takes the natural patterns and rhythms of growing and harvesting and intensifies them in space and time. The greenhouse creates its own ecology and is an everchanging microclimate. Plant life fosters animal and insect life and a system of interdependence emerges. You, as the gardener and caretaker, set the stage and define the parameters of this process.

The greenhouse gardener has some advantages and problems that the outdoor gardener does not have. Planting space is much more limited, for one thing. Also, an attached greenhouse is an integral part of your living environment; your home shares in the problems of the unit, as well as deriving benefits from it. The greenhouse gardener does have one great advantage over the outdoor gardener in not being as dependent on the weather. Low temperatures in the coldest part of the winter are the only exception. The ravages of wind, hail, drought, and frost have no effect on greenhouse plants, whereas outdoors they can and do destroy many crops. As a greenhouse gardener, you enjoy both a greatly extended and a much more predictable growing season, because you control the climate in your unit.

One of the most innovative aspects of the solar greenhouses in this book is that they are geared mainly to the production of vegetable food crops rather than flowers and houseplants. Food production and supplemental heat transform the greenhouse from a luxury item to a functional addition to the home. What you get from the greenhouse, though, is *directly proportional* to what you put into it. The heat is dependent upon the design and construction, and once the greenhouse is complete the temperature range is pretty much determined. But the food you get from the greenhouse changes continuously according to the season, the weather, and your efforts. A greenhouse will maintain itself with little input, but with regular care and attention it can perform beautifully. Four to five hours of work a week in a small, attached greenhouse can produce enough to provide a family of four and many guests with 80 percent of their fresh vegetables. This we know from experience.

Figure 85

A productive greenhouse calls for your care in a variety of ways. Five or ten minutes first thing in the morning is a good time to just look in on the greenhouse and see how it is faring, check the plants, and take count of the insects. This is also a very nice way to wake up. Then, at other times of the day and depending on the season, you will be doing various gardening tasks that will be described throughout this chapter.

We plant our greenhouses by the season, growing what grows naturally during different times of the year—although in the greenhouse these seasons are greatly extended. Also, we use only the light and heat of the sun to grow our crops, no artificial lights or supplemental heat. The garden is self-sufficient in these ways.

It's important to learn to work with the planting cycles in the greenhouse. Keeping written records, either a diary or charts (see Appendix I), is a helpful technique that takes no more than a couple of minutes a day. Beginning greenhouse gardeners looking for a good, reliable system may want to do what many farmers have always done: follow the phases of the moon in their planting through each season. We do this and find it very effective.

First Quarter (increasing from new moon to about half full)
Plant the vegetables whose foliage we eat. Examples are lettuce, cabbage, celery, spinach, etc.

Second Quarter (increasing from about half full to full moon)
Plant the vegetables whose fruit we eat. Examples include beans, peas, peppers, squash, tomatoes, etc. An exception to this rule is the cucumber, which seems to do best when planted during the first quarter.

Third Quarter (decreasing from full moon to about half full)
Plants the vegetables whose roots we eat, such as onions, radishes, carrots, etc.

Fourth Quarter (decreasing from about half full to new moon)
Best for cultivation, pulling weeds, destroying pests.

The planting of flowers breaks down into annuals during the first and second quarter, perennials and biennials during the third quarter.

Of course, the vegetables and flowers you select to plant in accordance with this lunar calendar are dictated by the season. You shouldn't plant tomatoes in December regardless of the phase of the moon. But by superimposing the lunar cycle onto the seasonal one, you can create an effective structure for maintaining a successful greenhouse garden.

Greenhouse Layout

The size, shape, and number of planting areas included in your greenhouse will be determined by the configuration of the unit and your own judgement (practical and aesthetic). However, since space is limited, plan the layout of planting areas with the utmost care. For maximum productivity, leave narrow walkways and put the rest of the floor area into beds and tables. Vertical and tiered space should be used to the greatest extent possible. Shelves lining solid walls, planters on top of fifty-five gallon drums, and hanging planters are some effective methods of using this space. Movable shelves on brackets along the clear walls work well to hold trays of seedings. By June these shelves can be taken down and stored (Fig. 86). Temporary tables can also be erected in beds for seedlings (Fig. 87). Another good way of using vertical space was devised by the University of Arizona Environmental Research Laboratory. It involves a series of hanging pots, attached one above the other (Fig. 88). This ingenious idea really makes the most of greenhouse growth/space potential.

Figure 86

Some of the space in your unit will have to be given up to equipment. If you already have a tool shed, then you won't need as much storage area in the greenhouse. But empty trays and pots, cartons for

Figure 87

seedlings, small hand tools, containers of fertilizer, and other basic necessities of greenhouse living do accumulate quickly. When planning the interior layout, leave space for equipment and a comfortable amount of standing room. An example of an interior layout is given on p. 85.

Soil

The greenhouse can be a successful year-round garden if the soil is kept in top condition. To be fertile, soil must contain three elements: minerals, air and organic matter. Gardeners used to ''taste'' their soil before planting. If it tasted sour or bitter, it was no good for raising plants; but if it tasted sweet, a high yield could be expected. The soil that tasted sour was too acid; soil that tasted bitter was too alkaline. A good balance between the acidity and alkalinity of soil is important for a good harvest; most plants like a neutral or slightly acid soil. It is not necessary to taste the soil, however; if you have questions about it, send a sample to a county agent or buy a do-it-yourself soil-testing kit.

A healthy plant needs nitrogen (N), phosphorus (P) and postassium (K) from the soil; trace mineral elements are also essential. Nitrogen is needed for growth, particularly for green leaves. Yellow leaves and poor growth denote a lack of nitrogen in the soil. Organic matter decaying to humus releases nitrogen slowly, but is a steady source. Cottonseed meal and blood meal are other excellent sources of nitrogen. Phosphorus is needed for roots, fruit development, and resistance to disease. It comes from organic decaying matter, bone meal, and from ground phosphate rock. Potassium is needed for the cell structure of the plant and is obtained from organic matter and wood ashes. Trace mineral elements come from a variety of components in decomposing organic matter.

Figure 88

Air is another vital ingredient in the composition of healthy, fertile soil and in producing well-rooted plants. Along with using organic matter, vermiculite, or sand, another way to aerate and allow water to penetrate the soil is by providing a good supply of earthworms. Worms also fertilize soil through their casts which are rich in nitrogen, potash, and phosphate, and help balance the composition of the soil; if your soil has a good number of earthworms in it, you do not need to add nutrients as often or worry as much about insects. There are over two thousand kinds of earthworms but the two that are of interest in the greenhouse are the soil-ingesting blue and the manure- and composting-ingesting red. Both make humus. Research is now being done that may introduce new or little known varieties of earthworms. Keep a lookout for the results.

Traditionally, sterilized soil is used in the greenhouse to avoid importing bugs and diseases. However, we have never done this due to two major practical problems. Soil is sterilized by heating it, but how do you get a ton of soil (the average amount needed for ground beds in a small greenhouse) into a home oven? The other solution, buying presterilized or potting soil, is prohibitively expensive in such quantities. If you have access to sterilized soil use it, but if not, don't let this be an obstacle to planting your greenhouse. When you start seedlings, though, use a sterile medium such as 1/2 peat moss and 1/2 vermiculite.

The soil in a ground bed should be 12-16 inches deep, and under it should be 4-6 inches of pumice, sand, vermiculite, or pea gravel to insure proper drainage. Greenhouse soil should *not be as dense* as regular garden soil and it ***must*** drain well. Plants do not like sitting in a puddle of water. If your soil is on the heavy side, add a good portion of vermiculite and organic matter. Two good soil mixtures are:

1/3 rich organic topsoil
1/3 peat or compost
1/3 vermiculite or perlite

OR

1 part of organic matter
1 part of sandy loam soil

If you live near the mountains, try this one:

1/4 arroyo sand (river bottom)
1/4 black mountain earth (found under the pines)
1/4 pumice, perlite, or vermiculite
1/4 compost or peat moss

The same soil can be used for a long time in a greenhouse, but it must be regularly enriched with compost and organic matter. By fertilizing, rotating plants, and using nitrogen fixers (such as peas) in various locations, I have kept the soil in my greenhouse productive for five years, replacing only the top couple of inches every year. (See sections on crop rotation, page 91, and fertilization below).

Carbon Dioxide

Together with sunlight and water, carbon dioxide is essential for photosynthesis. Minimum CO_2 for plant growth is .03%; doubling CO_2 in a greenhouse can approximately double photosynthesis. Increasing carbon dioxide levels in the greenhouse can partially offset the reduction of light in winter. Keeping animals (such as rabbits) in the greenhouse, evaporating dry ice, or having a functioning compost bin increases the carbon dioxide level. An attached greenhouse has a CO_2 advantage over an independent one in that the air from the home circulating through the greenhouse on winter days is naturally higher in CO_2 than outside air. The greenhouse, in turn, gives back to the home oxygen-rich purified air. This is another example of the symbiotic relationship between house and greenhouse.

Fertilizers

Every time a crop is harvested, the soil should be fertilized. Remember that you are asking a great deal more of the greenhouse earth than the garden earth in terms of length of the growing season and density of the crop. Because of the relatively limited space, most greenhouse owners plant intensively. We deliberately crowd our crops in order to squeeze every last tomato, bean, and pea from the unit. This type of intensive planting does work, if the soil is rich enough. Regular fertilization is the primary way to enrich the soil.

There are many different types of fertilizers. A local nursery can recommend a variety of basic dry and liquid commercial brands. If you apply any of them every time you are ready to plant a new crop, the soil should stay in good condition. Your plants are the best indicator of insufficient or incorrect fertilization. Liquid fertilizer is available to the plants immediately, while dry fertilizer takes six weeks to three months to be absorbed. If you wish to mix your own fertilizer, blood meal is an excellent source of nitrogen and some phosphate. Rock phosphate also supplies phosphates and wood ashes supply potassium. Well-composted manure provides many trace elements as well as being a source of nitrogen, phosphate, and potassium. Iron comes from blood meal and magnesium from seaweed or Epsom salts. I use a liquid fertilizer (fish emulsion) in my greenhouse every two or three weeks. I use dry fertilizers, cottonseed meal, bone meal, and blood meal each time I have harvested one crop and before I plant another.

A compost bin in the greenhouse is another valuable source of fertilizer. It supplies decayed organic matter, the best source of the essential mineral nutrients for the plants. And as the organic material is decomposing, it supplies heat and carbon dioxide to the greenhouse. The compost pile must of course be well planned and maintained so that it does not smell foul. This is not a problem if it is kept active with manure and green material (green leaves, plant stems, flowers, roots, lawn clippings, etc.) and layered with dry material (hay, straw, dry leaves, sawdust, etc.). Each layer should be from 2 to 4 inches in depth. When the green materials have lost their moisture, water should be sprinkled on the pile; apply only a little, as too much water will cool the pile down. Compost temperatures should be between 140°F and 160°F.

The biggest drawback to a compost pile in the greenhouse is the space it takes. The bin should have a floor space at least 3 feet by 3 feet. And you need to have enough room in front of the bin to comfortably turn the pile every 3 or 4 days. But if you have a greenhouse large enough to house a compost bin, the rewards are well worth it.

Figure 89

Other organic fertilizers, such as manure, work beautifully. Dry cow and horse manure are particularly effective. Keep in mind, though, that every time you bring foreign matter into the unit, you run the risk of bringing harmful insects with it.

Manure and compost should also be added to the soil while the plants are growing and producing. Frances Tyson makes what she calls "horse manure tea" by soaking fresh dung in a pail of water. Frances periodically pours this hearty brew on her soil; the health and beauty of her plants (see Fig. 89) testifies to its value as a fertilizer.

Many people believe that organic fertilizer is more effective overall than chemical fertilizer (we're among this group). We've used fish emulsion very successfully. I like the stuff, although it makes my whole house smell like Provincetown at low tide for a couple of hours. But what you use depends to a great extent on what is available. The most important thing is to fertilize regularly, lavishly, and deeply. When fertilizing, turn the soil over with a shovel or pitchfork. Dig down to the bottom and really move that earth around. You will bury the fertilizer deep in the soil, the best place for it. It will feed the whole root system of your plants and also encourage the roots to grow deeper, thus creating a firmly well-rooted crop.

Mulching will enrich the soil and keep it evenly moist. Mulch with organic substances that break down readily into humus. Grass clippings, leaves, wood shavings, vegetables, and other leftovers—anything you would compost—will work very well in the greenhouse. You can literally turn your beds into compost piles, using nearly all your organic garbage, mulch, and fertilizer on them. Keep a small jug by your sink and every day or so convert your "garbage" into your greenhouse soil.

Get an image in your mind of the soil as a living element. Nourish and care for it just as you do your plants. The soil essentially makes the difference between a sterile and fertile greenhouse (Fig. 90).

Figure 90

Hydroponics Versus Soil

Hydroponics is a method of highly controlled agriculture whereby plants are grown in a "non-soil" medium such as sand or vermiculite. A balanced diet of nutrients is fed directly to the root system.

There is a long standing battle raging between hydroponic and soil aficionados. Hydroponic gardeners claim as high as 300 percent increases in yield; soil users retort, "Hydroponic plants are chemical junkies and besides they don't taste good." I have limited experience with hydroponics but have found the following:

1) Hydroponic plants are more dependent on continuous care. A soil plant in a deep bed can survive several days in midsummer without watering. A hydroponic plant will die if a feeding is missed. For this reason, most hydroponic operations call for an electric timer, feed batch tank, tubes to plants or containers for feeding and drainage. In a small home operation these items are not particularly expensive or energy consuming.
2) I have seen plant growth and densities in soil that are not surpassed by any hydroponics techniques, as in the Tyson's greenhouse (Chapter VIII, p. 149 and Fig. 89). Granted, this is a small operation that receives a maximum of tender love and care; in a larger commercial unit increased yields by hydroponics are documented fact.

3) The soil in beds and on tables has the ability to store heat. Some hydroponic media such as pea gravel can also thermally charge but other media such as vermiculite do not hold much thermal storage capability.
4) For the complete novice I recommend starting with soil. The greenhouse is a new experience by itself. The added element of hydroponics might complicate things. After you have had some experience with the greenhouse, run a hydroponic experiment and then decide which side of the battle to join.

We have included some good hydroponic books in the reference list.

Straw Bale Culture

A compromise between growing in soil and hydroponics is to use straw as the growth medium, a method I learned about from Don Knight of Santa Fe. This is a standard commercial procedure in Great Britain. The advantages of this method are: quantities of soil do not have to be hauled into the new greenhouse; since straw is lighter than soil, it is a particularly appealing option for a second-story greenhouse such as the one in Chapter VIII (p. 108). The straw medium is constantly composting, providing both nutrients for the plants and gentle heat for their roots and for the greenhouse; composting straw produces CO_2; plants such as tomatoes enjoy being able to form large root systems in the loose structure of the straw. The space where you will put the bales must first be covered with a layer of plastic material; put bales bound with wire on top of the plastic. Next there is a three-week period of preparation. Soak each bale daily with one gallon of water for two weeks. To avoid evaporation, it is helpful to cover the straw after each watering with thin plastic. After the bales are thoroughly soaked (and composting of the straw is well underway) the next step is the application of fertilizer (see p. 82). The fertilizer must provide the following nutrients: nitrogen, phosphate, potassium, magnesium, and iron. Commercial tomato food contains all of these; the ratio of available nitrogen, phosphate and potassium is approximately 5:10:10. Two and one-half pounds of this fertilizer will be needed per bale of straw.

The initial application of fertilizer should be introduced into the straw by applying it dry to the top and then wetting it with one gallon of water per bale per day. If the straw appears too dense or the fertilizer at first too insoluble, work it into the straw with a hand tool. Getting the fertilizer to penetrate rapidly increases the bacterial action in the composting straw. The internal heat will rise to quite high temperatures. To hold this heat, continue to keep the bales covered. The composting straw is generating nutrients needed for the plants about to be placed in the bales. Within a week, the high temperatures generated in the straw will fall below 100°F. Pull out the sprouts that have come from the seeds in the straw. Then it's time to plant.

Plant 6- to 8-week old seedlings that have been transplanted at least twice. Put no more than 2 tomato plants in each bale. Surround every seedling with a ring of cardboard or metal 3–4 inches high and about 6 inches in diameter, and fill it in with rich soil or compost. If your plant will need to be staked, this is a good time to do it.

As the young plants begin to send their roots into the enriched straw, they will rapidly make use of the fertilizer present. It is necessary to continue feeding periodically. Beginning 2 weeks after planting and every other week after that, add about 1/4 of the amount of nitrogen fertilizer. On alternate weeks feed with potash, again about 1/4 of the initial amount. You can vary the water to include a gallon or two of manure tea or fish emulsion. With this schedule of feeding and watering (and training and pruning) your plants will grow well.

Planting Layout

Before planting, give careful thought to the right light and heat requirements, the shape and size of various crops. Seasonal light changes affect plant growth and photosynthesis, as does temperature. The sun follows different paths in the sky at different times of the year. At midwinter when it's lowest in the sky the sun will be high on the back wall of the greenhouse. In midsummer, the sun is at its highest and only the front of the greenhouse is in full sun. These changes cause different microclimates and will dictate to some extent where you plant certain crops at various times of the year.

Different areas of the greenhouse also create different growing environments. The higher areas will always be the warmer; the front will have full light, while the back will be shadier. During the winter, the back area will not experience as wide a range in temperature variation as the front area will. Learn the requirements of your plants and put them into the area of the greenhouse where they will be happiest.

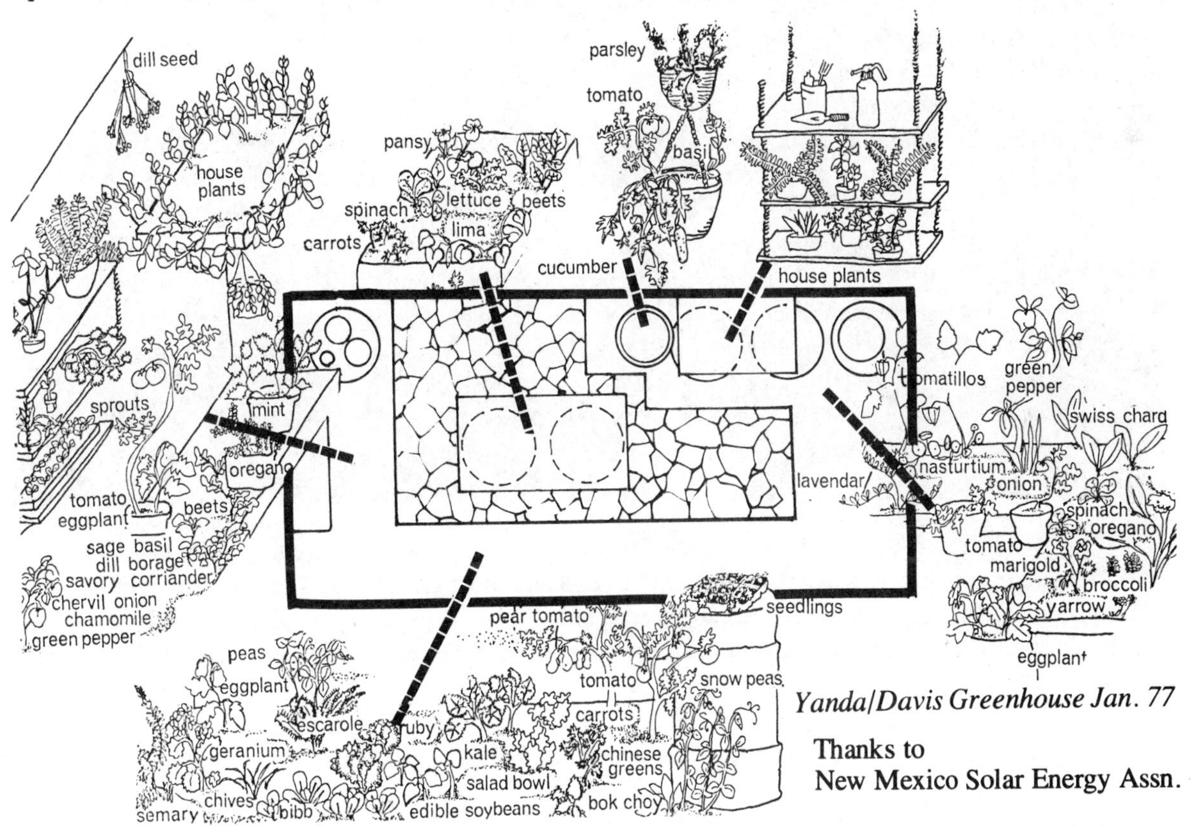

Yanda/Davis Greenhouse Jan. 77

Thanks to
New Mexico Solar Energy Assn.

Figure 91

Planting fruiters—eggplant, tomatoes, peppers—in the front of the greenhouse in spring, summer, and fall will allow them to receive the full sun they need. The greater challenge is finding plants that will grow in the back beds. Greens, root crops, legumes (peas, beans), and some herbs do well in these area. They prefer the cooler temperatures and can tolerate lower light levels than the fruiters.

You will probably want to experiment with a variety of planting layouts. Just remember to give extra attention to the plants in the rear. Provide them with as much light as possible or they may tend to become *phototropic*.

Phototropism is the tendency of plants to bend toward their source of light. In the outdoor "natural" growth area plants receive light from all sides, though much of that light may be from the south. In a greenhouse, this "diffuse" source of light may not be available, especially if the north wall of the solar greenhouse is dark in color for heat storage. In such a case, the plants will bend toward the south glazing.

This bending is caused by an accumulation of growth-regulating *auxins* along the "dark" side of the plant stem, causing it to become distended or bent. Wasted plant energy results, though plant growth is not greatly reduced. Using glazings that diffuse sunlight, and painting interior, nonmassive, walls white, will more evenly distribute light to the plants.

Figure 92

As you get ready to sow your seeds, keep in mind that most greenhouse gardeners plant intensively (see Fig. 92). The number of plants you can squeeze into one bed depends a great deal on the fertility of the soil. Under good conditions you can cut the recommended spacing by ¾ths for many crops and get a fine harvest.

The Vegetable Planting Guide

Don't grow any vegetables that you don't enjoy eating. It's no major accomplishment to grow radishes or the hardy cold-weather crops in your solar greenhouse at any time during the year. But if you dislike radishes then it's a waste of time and soil nutrients. Raise the vegetables that you and your family eat and your success will be measured not only in terms of what you can grow but what you use.

Having said this, we encourage you to plant any vegetables or flowers that you especially like. If you take proper care of the plants, start them at the right time of the year, ward off insects (see p. 92), and follow the other basic procedures we are suggesting, there is no reason why you can't grow almost anything you want to. We urge you to experiment, particularly if you enjoy gardening enough to take a few risks in return for possibly spectacular results. Imagine how you would like your greenhouse to look and then simply bring that vision into reality. It can be done.

Thus far we have talked mainly about vegetables, but don't feel compelled to use every square inch of the greenhouse for edible plants. Flowers add color, fragrance, and beauty to your unit year-round. If the greenhouse space is an attractive one, you'll find that you want to spend time in it and maintaining the plants will be a pleasure. The plants will enjoy your company and benefit from your care.

The greenhouse experiences seasons just as the outdoor environment does, but summer comes early and lingers on in the greenhouse garden. The planting cycle in the greenhouse is based on this extended growing season. Although the planting cycles will vary regionally, the guidelines we will give you apply to most areas that experience moderate to great seasonal variation—including, most importantly, cold winters. In these regions the greenhouse seasons can be generally defined as follows: **spring** is February through April; **summer** is May through September; **fall** is October through November; **winter** is December through January. Some crops will be harvested within one period, others will extend through two seasons, and still others, notably tomatoes, will continue to bear through three cycles.

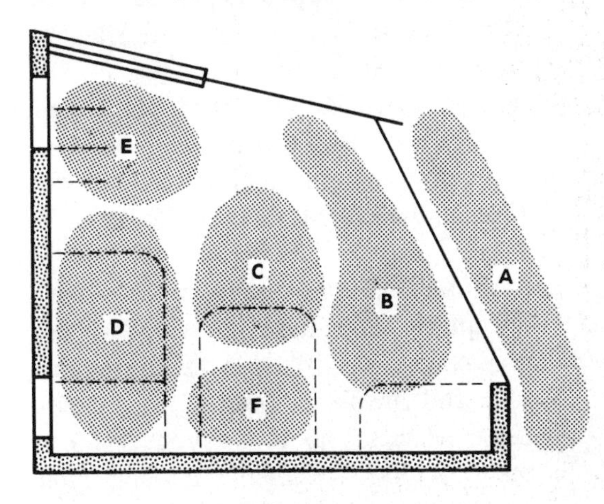

Figure 93

Profile: SPRING—SUMMER—FALL

A. Shading plants
B. Fruiters. Tomatoes, cucumbers trained up twine. Trim foliage, squash, melons.
C. Seedlings, herbs, fruiters. Hydroponic table.
D. Low light, coolest greens. In late summer new fruiters can go here. Climbers, flowers.
E. Flowers, shade lovers.
F. Berries, shade lovers.

Spring. While there is still snow on the ground and the temperature outside says winter is here, it is wonderful to experience spring in the greenhouse. In February start seedlings in trays for the tomatoes, eggplant, and herbs that will be planted in the greenhouse to produce through summer and fall. The hardy cold-weather crops such as peas, leafy greens, onions, radishes, and broccoli can be planted directly in the greenhouse beds. Even though you may still have these cold-weather crops in the beds from the winter cycle, they are usually tired after battling the cold and in most cases you will want to remove them or turn them under for compost. The spring crops come up quickly and are healthy and strong.

By the middle of March all crops are doing well. At this time transplant the seedlings in trays into the greenhouse beds. Interplant everything that grows well together as this helps confuse the insects and allows you to plant more intensively (see Companion Planting, p. 91).

The middle of March is a good time to plant melons, pumpkins, cucumbers, and any other shallow-rooted (less hardy) fruit-bearing crops you want to keep in the greenhouse through the summer. Plant melons near a low vent so that you can grow them out through this opening when they are large and the weather is warm enough to have the vent open all the time. Many people are tempted to start their outdoor garden seedlings at this time, as everything in the greenhouse is growing like mad. Do not! It is too early; garden seedlings will become rootbound and too big to

Figure 94

transplant, unless you are lucky enough to live in an area where the last freeze is the end of April. Pole beans do well when started now and are a very successful greenhouse crop (Fig. 94). They put most of their energy into vertical growth and thus do not take up much space in relation to the amount of beans they produce. You do have to provide a lattice work of stakes for them to climb, and it should extend to the ceiling, as the beans will easily reach that height.

By April the greenhouse is in full swing. This is when I begin giving away fresh vegetables to any willing neighbor, since there's always more than we can eat. During the middle of April, begin planting seedlings for the outdoor garden. This is also a good time to start plants that will be planted just outside the south face of the greenhouse for shading (Fig. 95). Sunflowers, morning glories, and jerusalem artichokes all work well. Grapes can also be planted in this area. Train the vines up and across the front face; in the fall cut them back.

Figure 95

Summer. During the summer season the plants you have growing in the greenhouse will flourish, with regular care and attention. Fruiting vegetables begin producing about two months before their outdoor counterparts, and you should have a steady harvest throughout this season. By the second week of May all the seedlings you started for your outdoor garden are well established. *Harden off* garden transplants by exposing them to harsher outdoor weather for about two weeks. In the beginning, bring them in at night. Also, some vegetables that were planted directly in the beds can be moved outdoors even though they are quite large by this time. Tomatoes, peppers, and broccoli, blossoming and fruiting, have been transplanted with no ill effects on their productivity. My broccoli actually benefited from the move as the garden was cooler than the greenhouse.

Keep the greenhouse well ventilated. It's a good idea to close the greenhouse off from the house and open vents and doors to the outside. This, however, may bring unwanted pests such as grasshoppers, chickens, dogs, and cats in, so you would be well advised to put chicken wire over low vents and a folding gate on the door that will keep out the larger pests. If you put screens up you must increase your vent size substantially as screens decrease the air flow necessary for good ventilation. The greenhouse must always be kept clean and the population of harmful insects under control, but during this time of heavy production, it is especially important. Train climbing vegetables up lattices, string, or stakes, and keep foliage trimmed.

At the end of August begin seedlings in trays for the crops you want in your fall and winter greenhouse as they must be well established and do most of their growing before the cold nights set in. It's a good time to start a few of the smaller tomato varieties.

Fall. Early October is a good time to thoroughly clean the greenhouse and fertilize the soil. The plants that grow through late fall and winter need all of the help they can get. Dig out the top 2-3 inches of unplanted soil and replace them with compost, or fertilize well using another method you prefer. Also dig around the plants that will continue to grow in the greenhouse to loosen the soil, and fertilize them deeply before you transplant the seedlings you started in August.

Leafy greens, broccoli, cabbage, peas, onions, root vegetables, and herbs all do well in the fall greenhouse (Fig. 96). Tomatoes, beans, and eggplant continue to produce, but usually cucumbers, melons and peppers are past their prime.

Figure 96

Winter. Because of the lower temperatures and the shorter daylight hours all plants show a slow growth rate in the winter greenhouse. You may find this is a good time to *really* clean out the unit; wash *all* the containers and boxes; replenish the soil. This is an excellent time to get rid of any lingering pests by freezing them to death. Simply open up the greenhouse at night for a week to ten days. Close it during the day and turn the soil over several times. It is a valid decision to give your greenhouse and yourself a rest during this time. On the other hand, you may find it the most rewarding time of year to have a garden. Hardy and cold-weather plants will grow during this period without added heat. I love the nightly fresh salads and herbs and the weekly serving of fresh broccoli and peas we eat through this period.

The planting chart (Appendix I) will give you a full listing of varieties, time of planting, and locations we've used in our greenhouses.

Profile: WINTER

A. Lightest, coldest. Leafy greens, radishes, peas, broccoli, roots, tubers. Carry over fruiters.
B. Light, cool. Herbs, greens, flowers. Transplant seedlings.
C. Light, warm. Hanging pots, flowers, herbs.
D. Light, warm. Winter tomatoes, peppers. Climbers, beans, houseplants.
E. Low light, warmest. Start seeds, sprouts. (On shelves, bread will rise well here.)
F. Shady, cool. Berries.

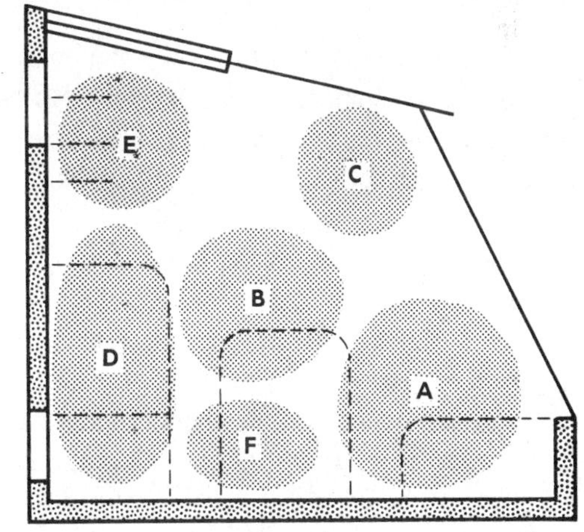

Figure 97

Flowers in the Greenhouse

Houseplants. All houseplants seem to thrive in the solar greenhouse. Many people give their plants an invigorating outdoor vacation during the summer. A vacation in the greenhouse is even better. The color, size and density of foliage improves tremendously, and the plants look as good or better than when they were nursery-new. Therefore, any plant suffering from the windowsill, coffee table, or bookcase blues should be transferred into the greenhouse for a couple of months or more.

Light and humidity in the greenhouse are two important factors in reviving plants. Be sure, though, to place plants that are susceptible to burning in a location away from direct sunlight, just as you would in your house.

Hardy houseplants can stay in the greenhouse year-round, unless you are planning a solid winter freezeout. Geraniums are one of the best examples of this. Once, at about 22°F, mine died down to the roots, then made a magnificent comeback in the spring. If your greenhouse temperatures never drop that low, cut back the geraniums in midwinter anyway. By April they will be beautiful again.

Strawberry begonias have also proven to be very frost-resistant houseplants; in fact, they seem to thrive at cooler temperatures. Wandering jews are remarkably hardy, though again, they die back when the temperatures are in the twenties. The color and vigor of the new leaves in spring is incredible.

More tender houseplants, such as the coleus, benefit greatly from the greenhouse environment, but they must be brought back into the house during the coldest months unless your unit never freezes. In this category, fuchsias, swedish ivy, and spider plants are especially content in the greenhouse. Treat them exactly as you do in the house in terms of watering, trimming, and feeding.

All the flowering plants you normally have in your garden grow beautifully in the greenhouse. Choosing those to include in the unit involves considerations of space, size, and general planting layout. Smaller varieties are often preferable because they add color but don't require much space. A favorite in this category and probably one of the easiest is the petunia. Petunias come in a wide range of colors and bloom profusely. If they get spindly, cut them back, and you'll see them return in healthy vigorous condition. Petunias are considered tender annuals, but in the greenhouse they can become perennials (hardy to about 22°F). The same applies to marigolds, nasturtiums, and pansies. Pansies, like petunias, need to be cut back periodically. All these types of flowers bloom continuously.

If you have a dark moist corner in your unit where nothing seems to grow, try ivy. Buy a variety that is recommended for shady north-facing walls and your problem corner will be transformed into a haven of green.

Gladiolus bulbs planted along the border of the front beds grow very well. In many regions gladiolus are not reliably hardy, and bulbs planted in the garden have to be dug up and brought in during the winter. In the greenhouse, however, they remain in the ground year-round; they die back to the bulbs in late fall, proliferate, and return with increased beauty in the spring.

Transplanting. The greenhouse provides prime conditions for starting flowers to transplant into the garden. You can get a good headstart on the outdoor growing season and thus get much more enjoyment from your flowers during the summer.

In early April start perennials such as poppies, daisies, and chrysanthemums in the greenhouse; they can be transplanted sooner than annuals and will usually bloom the first year with this kind of headstart. Plant annuals, including marigolds, zinnias, morning glories, and sweet peas in mid-April.

One of the advantages to starting flowers this way is that April temperatures in the greenhouses are usually warmer and more consistent than outdoor temperatures in early June. Seeds therefore germinate faster; also, the soil can be kept evenly moist more easily during the crucial time of germination, and this improves the success rate for sprouting seeds. You can also set the plants out, already several inches high and hardy, in any configuration you like. They are properly spaced and will remain that way. I have never been able to achieve this kind of symmetry (without a lot of effort and frustration) when sowing seeds directly in the garden beds.

Many flowers can also be transplanted into the greenhouse in fall. If you are careful to get all or most of the roots, they will continue blooming all winter. Petunias, pansies, and marigolds are the three I've transplanted successfully.

Generally it doesn't make much sense to dig up hardy perennials that are well established in the garden, but you should follow your own preferences in deciding what flowers to transplant. When you do get ready to bring them in, dig up the roots with plenty of soil around them. This *lessens* the shock of transplanting, decreases the danger of damaging the roots, and gives the plants the illusion that nothing unusual has happened to them. By the time their roots have grown beyond that ball of soil, they should be firmly grounded and ready to cope with any differences in the growing conditions.

Maintenance

Watering. The amount of watering that greenhouse plants need varies according to the season and climate. In summer and over prolonged periods of clear days (any time of the year) they will require more water than on cloudy days and during most of the winter. Determine the amount of watering you will need based on personal observation and experience. The best way is to stick your index finger down into the soil. The soil should be moist, not slushy.

DO NOT OVERWATER! A greenhouse uses much less water than an outdoor garden. Humidity should not be a problem in the greenhouse if you run it properly. Overwatering can rot the roots and create other problems. Use rainwater whenever possible; it is good for the plants.

In warm weather the best time to water the greenhouse is about noon. In summer evaporating water cools off the unit and keeps the temperatures comfortable during the afternoon. In winter a morning watering with lukewarm water will help the plants after a cold night. During the cold months it is best not to water on cloudy days.

The equipment you use can be as simple as a large watering can. In the summer you can water the greenhouse with a hose from outdoors. Ideally, though, install a hose in the greenhouse for year-round use, with an inside faucet to attach it to. Incorporating such an arrangement into the design of your greenhouse will save time and work. It is also helpful to have attachments for your hose that will allow you to spray a fine mist, beneficial for newly planted seeds and young seedlings.

Slow watering is always preferable to quick, forceful watering. The soil is better able to absorb the moisture and you'll get a true idea how much the ground needs and can use. A quick flood will not penetrate as deeply to the roots and more is lost through evaporation.

Crop Rotation. Rotating crops is a highly recommended method of maintaining healthy soil and getting a good harvest over a number of years. Different crops take varying amounts of nutrients out of the soil; some actually enrich it.

The "heavy feeders" include tomatoes, cabbage (and members of that family), corn, all leaf vegetables such as chard, lettuce, endive, spinach, and celery, cucumbers, and squash. They need to be well fertilized and should not be planted year after year in the same bed.

Legumes (notably peas and beans, including soybeans and lima beans) are good soil builders; they "fix" nitrogen from the air, storing it in the soil for plant use. By alternating crops of legumes with the heavy feeders, you will be helping the soil to maintain a healthy balance of vital elements.

A third category consists of crops called "light feeders." Herbs and root vegetables belong in this group. If you're uncertain about how to classify a plant, it's usually safe to include it here.

Flowers break down into these same divisions. But unless you put all of the beds into flowers, they should not present a big problem. Just relocate them from time to time and keep the soil well fertilized.

Companion Planting. Vegetables and flowers grown side by side complement each other and make the greenhouse more attractive. Companion planting, mixing one plant with another in the vegetable row or flower bed, is a good practice in any garden but is especially beneficial in the greenhouse.

Some reasons for companion planting are: certain plants excrete a substance above and/or below the ground that protects the plant (other plants growing nearby also benefit). A large plant that needs a lot of light may be a good companion to one that needs partial shade; a deep-rooted plant may benefit a shallow rooted plant by bringing up nutrients from deeper soil layers. Companion planting increases diversity in the greenhouse and tends to frustrate insect feeding and thus thwart buildups of insect populations. Plant clumps of herbs such as basil, anise, and coriander throughout the greenhouse. Garlic and marigolds are good companions to have in every bed. In fact, garlic is such an effective insect repellent that it's a good idea to keep a refrigerated jug of water with crushed garlic in it on hand to spray any whitefly, cabbage moth, aphid, or other harmful insect that dares enter your greenhouse.

In planning the layout of your crops, give some thought to companion planting. There are obvious mixes that will not work: two plants that grow tall, two that have deep roots, and so on. Some plants do not like others; for example, kohlrabi and fennel should not be planted next to tomatoes. Some plants that do well together are carrots and peas, kohlrabi and beets, tomatoes and parsley, celery and bush beans, radishes and lettuce, turnips and peas, onions and beans. Aromatic herbs make good companions as border plants.

Air Circulation. Plants need fresh air to replenish their oxygen for respiration and carbon dioxide for photosynthesis. If there is not sufficient air movement in the greenhouse, a layer of stale air surrounds each leaf and slows growth. Air movement through ventilation is the most effective way to control humidity and temperature, and thus to promote the propagation of vigorous, disease-resistant plants. In the summer air circulation is improved by having the exterior door and all exterior vents open. On winter days open vents to the house. Always make sure you have sufficient air circulation through the greenhouse.

Pollination. Your help may be needed as a pollinator in the greenhouse—the job may not be done effectively by wind and insects alone, as it is outdoors. There are two types of blossoms you need to become familiar with in order to pollinate correctly. The *complete* flowers, found on peppers, eggplants, and tomatoes, are pollinated by gently flicking the base of the blossom cluster with your finger. Melons, cucumbers, and squash have *incomplete* flowers; the male and female parts are on separate flowers and must be cross-pollinated. The female part is easy to identify as it has the immature fruit below the petals; yellow pollen is on a center stalk in the male blossom. Gently touch the pollen with a paintbrush or feather and transfer it to the female blossom. Pollinate daily and do it during the cooler part of the day; I have found early morning the best. Also, look for self-pollinating varieties in seed catalogs.

Dogs and Cats

Pets in the greenhouse can be a nuisance. Dogs with their bounding energy can trample seeds and plants into oblivion. They also love to dig into and lie down in a moist bed on a hot summer day. Cats consider the greenhouse beds a custom-made litter box. In general, all the problems you are likely to have with dogs and cats in an unfenced garden exist in the greenhouse. A gate is usually sufficient to prevent animal damage, but use your ingenuity if further barriers are needed.

Insects

Joan Loitz, owner of The Herb Shop and Solar Greenhouse in Santa Fe, New Mexico has prepared a "Commonsense Pest Control for the Home Greenhouse" manual that we are reproducing here:

COMMONSENSE PEST CONTROL FOR THE HOME GREENHOUSE

By Joan Loitz

Greenhouses are special places—good for you, great for plants, but equally suitable for undesirable pests. Have no doubt that in time they will find you. You can build a greenhouse in the middle of the desert and one day the slugs will appear. You can have a most tidy greenhouse, but a summer breeze may still carry aphids or whitefly in to take up residence on your tomatoes.

What to do?

- PREVENTION IS THE BEST CURE.
- IDENTIFY EXACTLY WHAT THE PROBLEM IS.
- TAKE ACTION.

Let's look at these three steps in greater detail.

Prevention is the Best Cure

Keep it Clean. Periodically remove dead and decaying matter from plants. Also, if you have entire plants that are seriously infected by disease or insects, get rid of them. Despite our occasional fond attachment to greenhouse plant "pets," it's always better to start new plants rather than risk the serious problem of spreading a major infestation or infection throughout the greenhouse. Clean matter (not infested by disease or insects) can be placed in the compost pile. Infested matter should be burned or thrown out in the trash. Never allow infested matter to propagate by burying or composting it. Typical compost piles do not reach high enough temperatures to completely destroy disease or insects.

Weeds are trouble. Weeds either in the greenhouse or within the immediate vicinity outside the greenhouse provide a handy breeding area for plant pests and diseases. By removing weeds from outside the greenhouse area in particular, you deprive insect pests of a secure home for overwintering their eggs. This reduces their chances of infiltrating your greenhouse during the winter as well as the next spring.

Beware of taking in boarders. Quite often greenhouse gardeners like to bring in several plants from their outdoor vegetable or ornamental gardens. Before doing so, each plant must be checked carefully. Outdoor plants can carry aphids or whitefly, for example. Bring in only the cleanest and healthiest plants. Minor infestations can be treated, observed, and then, only when the plants are clean, brought into the greenhouse.

Don't take in a friend's plants. Occasionally vacationing friends or acquaintances with soulful expressions and "sick" plants will seek the rehabilitative environment of your greenhouse. Thank them for their sincere confidence in your abilities as a plant sitter and healer and politely say, "no." Remember, whatever problems their plants have will compound with interest in your greenhouse.

Always buy plants and supplies from a reputable source; buy from a supplier who can answer your questions both now and, should you run into difficulty, later. When you purchase new plants, isolate them for awhile. Check for any bugs or other problems that could spread undetected.

Bleach it, boil it. We've discussed how to get rid of undesirable plant material, but what of the pot that contained it and the potential harm that it might harbor? With a little bleach or hot water containers can be made ready for planting anew with no danger of spreading anything undesirable.

Plastic pots, trays, cell packs, even wooden seed flats can be soaked in a 10 percent bleach solution (1 part bleach to 9 parts water). Soak for one hour, scrub if necessary, dry, and they're ready to use.

Also, twice a year (spring and fall) remove all pots from any wooden benches in the greenhouse, and scrub the wooden surfaces to remove debris; let them dry, then apply straight bleach to the benches, making sure to get it into all the cracks and crevices. Allow them to dry thoroughly in full sunlight for several days. Be sure to maintain adequate ventilation during this time and be careful not to splash bleach on plants growing under the benches. This process tends to discourage scavengers such as sow bugs, millipedes, and slugs that live in dead and decaying matter and the rotting wood bench.

Clay pots may harbor bacteria, fungi, and insect eggs due to their high porosity. Wash off any dirt, then soak them in boiling water for 30 minutes. Another trick taught to me by a friend is to wash dirt off pots, then put them in the dishwasher on hot cycle. (This is not advisable for plastic pots which may melt.)

Keep it moving. As winter approaches and temperatures drop, doors and vents are closed to prevent heat loss. Solar greenhouses in particular are efficiently designed to reduce this factor as well as inhibit cold-air infiltration. This is all well and good from the solar engineering viewpoint, but an ''airtight'' greenhouse will drive an experienced horticulturalist straight up the trellis. Why? An airtight greenhouse invariably produces an excessively moist environment of stagnant air, in which:

- Bugs thrive! Most plant pests and diseases love this kind of environment.
- Plants can starve for carbon dioxide. CO_2 is food for plants. In a stagnant environment, plants quickly use the available CO_2 and growth will slow substantially without a supply of fresh air.
- Plants will produce excessively tender and fleshy growth, and quite often they become incapable of even supporting their own weight.

Thus, one of the most effective deterrents of plant diseases and insect pests, and inexpensive promotors of plant growth and vigor, is simply FRESH AIR.

On sunny days, open vents and let in some air. For greenhouses attached to residences, keep doors and/or windows open between house and greenhouse to permit full air exchange. On overcast days, open a vent just slightly or for a short time. You might lose a little heat but you will make up for it in healthy plants.

Commercial greenhouses have large automated fans for circulating air. Home greenhouses can be ventilated with small room fans. One small greenhouse I had years ago was adequately ventilated with a small cooling fan from a computer I purchased at a local surplus store for $3. Use your imagination, but remember to *keep it moving*.

Watering. Excessive moisture in a greenhouse can cause a lot of trouble. The combination of warm days and cool nights particularly during the fall and spring can encourage many plant pests and diseases. Water well early in the morning on sunny days. Plants will need the water as they will dry out more rapidly and the sun and fresh air will insure the leaves are dry by sunset, thereby reducing humidity during the cool nights. Skip watering if at all possible on cold, cloudy days.

Keep them healthy. Healthy plants will stave off disease and pests far more successfully than plants that have been stressed by conditions inadequate for proper growth. This is true for any living organism. Provide the proper cultural conditions for your crops and you will be rewarded with increased yield and fewer problems.

Keep your eyes open. Make a habit of checking your greenhouse regularly for insect pests and diseases. A weekly 15-minute safari through your greenhouse with magnifying lens will help keep you on top of any potential problems.

Identify the problem. Is it a bug? Or is it a disease or a culture problem? If it is a bug, make sure you can identify it *specifically* (see chart on p. 99). If you cannot identify it, take it to a local nursery for identification or check with your local county agricultural extension agent.

Recently a friend arrived at my greenhouse carrying a jar of bugs. He was upset, as they were occurring in epidemic proportions in his greenhouse, and he wanted to know what he could do to get rid of them. Well, they turned out to be Junebugs that were attracted to the sap coming out of the new wood on their recently constructed greenhouse. Within a short time, the bugs left of their own accord, without ever touching the plants. All insects you encounter in your greenhouse will not be so harmless, but it serves to point out that specific identification can save you a lot of trouble. Certain treatments will be more effective on one type of insect than another. Moreover, biological controls (which will be discussed later) imply very specific knowledge of pests and the insects used to control them. So, it is very important to know *exactly* what you are dealing with right from the start.

Figure 98

Take Action Immediately

When you spot undesirables, what are your choices?

A. The first option is *topical treatment*—sprays and dusts. It is important to note the relative effectiveness of sprays versus dusts. Most greenhouse pests such as whitefly, aphids, mealy bug, scale mites, and others are *sucking* insects and are most effectively treated topically by *wettable sprays,* sprays that can be mixed with water and applied as a mist. Larger insects more common to the outdoor garden and occasionally to the greenhouse such as squash bugs, beetles and others are classified as chewing insects and are more effectively controlled topically with *dusts*.

(1) **Inorganic.** These are earthen materials such as lead or sulphur.

(2) **Botanicals.** These are naturally occurring pesticides made from plants or plant extractions. Pyrethrin and Rotenone are two good examples. They do the job but biodegrade rapidly and have little residual effect; they are very suitable for use on edible plants.

Note: Nicotine is also considered a botanical but is not recommended for home greenhouse use due to its extreme toxicity. Commercially it is available as nicotine sulphate and is very potent. Home preparations of nicotine made from tobacco leaves are less toxic but run the risk of spreading tobacco mosaic virus particularly toxic to tomato plants. For this reason, also, do not allow smokers to light up in your greenhouse or touch tomatoes.

(3) **Microbial.** *Bacillus thuringiensis* is the most widely used and successful microbial preparation, commercially marketed under various names, such as Dipel, Thuricide, and Biotrol. It is a bacterial that is very effective against members of the caterpillar family but is harmless to humans and animals.

(4) **Home remedies.** Many gardeners have devised special "home brews" featuring everything from garlic juice to hot pepper sauce to combat insect pests. The most effective concoction I've ever used is simply soap and water. Mix up a mild solution of soap (not detergent) and spray on plants until all surfaces are covered. This is very useful against small infestations of aphids.

(5) **Synthetics.** These are chemical preparations generally classified as chlorinated hydrocarbons (such as chlordane, DDT) or organo-phosphates (such as malathion, TEPP and diazinon). These usually produce quick results but do not degrade rapidly and have a high residual value. Most of these chemicals are extremely toxic and are not recommended for home greenhouse use, particularly those greenhouses that are attached to residences. Systemic insecticides are chemicals that are mixed with water and then fed to the plant. The poison then permeates the entire plant. When pests take a bite of the plant, they are thus poisoned and die. This is used most often on ornamental plants outdoors; it is exceedingly dangerous to use this in home greenhouses where any edible plants are being grown.

To apply sprays or dusts, use hand sprayers (like an old window cleaner bottle) for small jobs. One- to two-gallon pressure-pump sprayers are better for larger jobs. Dunking is also very good for pests that are good at hiding, such as scale and mealy bug.

Figure 99

Caution: Remember, all pesticides, botanicals as well as synthetics, are *poisons*!

- Use only the recommended dosage or less.
- Store in a cool, dark place away from children.
- Always wear protective clothing, gloves, and a professional mask or agricultural respirator when spraying. (Respirators are available from feed stores, local agricultural suppliers, and mail order houses.) A paint mask or simple bandana across the face is not sufficient.
- For treating greenhouses attached to residences, be sure the greenhouse can be completely sealed off from the house. Any leaks could allow harmful fumes to seep into the home.
- Keep everyone (including pets) out of the greenhouse during spraying and for a safe period thereafter. It is always best to spray late in the day or right before sunset when the greenhouse can then be closed completely and allowed to "cook" overnight. Ventilate thoroughly with fresh air in the morning before entering.

B. Mechanical methods of pest control:

(1) Whitefly love the color yellow. By hanging common yellow flypaper next to tomato plants (a whitefly favorite), then shaking the plant, you can trap quite a few.

(2) Cold fumigation. Most plant pests cannot survive a week without food. Thus, if you pick an appropriate time when there might be a lag in your planting schedule during the winter, just open the doors and let the greenhouse freeze-out for a week.

(3) The thumb and forefinger squash. (Use gloves if you're squeamish!)
(4) Vacuum cleaner inhalation. (Good on whitefly!)

C. The best solution for home greenhouse pest problems is biological control—employing bugs that like to eat other bugs instead of your plants for lunch!

Biological control attempts to preserve, introduce, or otherwise manipulate beneficial insects in order to control harmful insects whose numbers have gone out of control and are causing marked economic or aesthetic damage. There are various types of beneficial insects. These include broad-spectrum predators and parasites: ladybugs, praying mantises, green lacewings, Trichogramma wasps.

Most outdoor gardeners will be familiar with the valuable roles these insects play in controlling plant pests. But they only have limited success in a greenhouse situation. They are called broad spectrum because they like a little of this and a little of that. Also, they are highly mobile. They can be successfully used in larger greenhouses where a variety of insect life is available, but will tend to wander out of smaller greenhouses quite rapidly in search of more food.

Host-specific predators and parasites (see chart). There are very specific predators and parasites for spider mite, whitefly, scale, mealy bugs, and flies. In general, they are quite small and prefer only one or two types of insect pests to eat. They are not as mobile and thus very effective in greenhouse situations.

Maintenance of predators and parasites. Let's take a simple example to demonstrate how this process can work. You start off with a high number of insect pests (called the host), then introduce a small number of predators. The predators will begin eating the host and will then reproduce rapidly due to this plentiful food supply. In short order, you will have a large number of hungry predators with few host left to eat. At this point, depending on the species involved, a number of the predators will die or go dormant due to the lack of food if they are left on their own. The host will rise slightly in number and hopefully, with good management, a population will evolve containing a small number of host and predators. Should the host population rise again out of proportion, a few predators could be periodically introduced.

Another option includes maintaining a 'sacrifice' plant on which an infestation of a particular pest is allowed to continue separate from the main greenhouse. Periodically a few leaves are removed from the infested plant and placed as food in areas where predators are active but the host population is reduced.

The key to success in pest control by biological means is accurate, periodic observation, identification, and manipulation (when necessary) of respective populations of beneficial and harmful insects. For the home greenhouse gardener, this need not be elaborate nor complicated. Armed with an inexpensive 10-power lens and a little notebook, a weekly inspection of the greenhouse provides sufficient information to determine the status of the insect population.

In general, biological control should not be considered an instant cure but rather an investment that can provide effective pest control over a longer time period.

Biological control is an attitude as much as it is a technique. Many greenhouse gardeners, especially new ones, will prefer the convenience and fast results of commercial botanical and synthetic products for pest control. But there is an emerging segment of gardeners who are seeking alternatives that will yield effective results with less or no danger to their own health or personal environment. These alternatives are in keeping with a sense of stewardship and responsibility to the larger environment as well.

It is important to note that the use of predators and parasites for pest control was very widespread in the 1920s and 1930s in England and Canada. With World War II came DDT and many other petrochemicals; activity in the area of biological control was largely abandoned.

We know more now. The facts are more than sufficient to indicate that the widespread uses and abuses of highly toxic, residual pesticides have been carried out at peril to our own lives and environment. Moreover, many insect species and mites have proven most successful in producing mutants resistant to chemical treatment. Another phenomenon common to pesticide abuse is ''secondary emergence.'' You may have noticed that after treating plants for a bug problem with highly toxic residual pesticides everything seems fine until, all of a sudden, the same problem returns only worse. In treating the plants to rid them of unwanted pests you have also killed beneficial predatory or parasitic insects who were trying to do the work for you. Thus the basic problem has only been increased.

Biological control offers viable alternatives to these problems and presents a method of gardening with nature. The attempt is toward participation, not annihilation. You have to pay close attention to the greenhouse environment and do some homework, but the rewards of maintaining biological control are well worth working toward—and besides, the process is fascinating!

SOURCES OF SUPPLY

Biological Control:
Biotactics, Inc. (predators for spider mite)
22412 Pico Street
Colton, CA 92324

Rincon-Vitova Insectaries, Inc. (Lacewings, Ladybugs, Trichogramma, Fly parasites, predators for spider mite, mealy bug, whitefly)
P.O. Box 95
Oak View, CA 93022

Bio-Control Co. (Trichogramma, Ladybugs, Praying mantis)
10180 Ladybird Drive
Auburn, CA 95603

SOME GOOD REFERENCES:

The Gardeners's Bug Book (4th edition)
by Cynthia Westcott, Doubleday & Co., Inc., N.Y.

Organic Plant Protection
Edited by Roger B. Yepsen, Jr.
Rodale Press, Inc., Emmaus, Pennsylvania

Windowsill Ecology
by William H. Jordan, Jr.
Rodale Press, Inc., Emmaus, Pennsylvania

* * * * * * * * * * * *

Common Pests: Description and Treatment

PEST	DESCRIPTION OF PEST	DESCRIPTION OF PLANT DAMAGE	TOPICAL TREATMENTS	NATURAL PREDATOR OR PARASITE	REMARKS
WHITE FLY	1/8" LONG, SMALL WHITE: WINGED. QUITE OFTEN PRESENT IN LARGE NUMBERS ON THE UNDERSIDE OF LEAVES, WHEN THE PLANT IS DISTURBED THEY FLY OFF IN A CLOUD.	WHITE FLY SUCK PLANT JUICES AND CAUSE LEAVES TO YELLOW AND DROP OFF	PYRETHRIN ROTENONE REPEAT WEEKLY UNTIL CLEAR	ENCARSIA FORMOSA (PARASITE) EFFECTIVE IN GREENHOUSES	WHITE FLY WILL REPRODUCE MORE RAPIDLY WHEN NIGHTTIME TEMPERATURES ARE LOW. WATCH OUT FOR ANTS.
APHIDS	1/16" LONG. SMALL, SOFT BODIED, PEAR SHAPED SUCKING INSECTS. CAN BE BLACK, GREEN, YELLOW, BROWN OR GREYISH IN COLOR	THEY CONGREGATE ON THE YOUNG SUCCULENT GROWING TIPS AND FLOWER BUDS CAUSING STUNTING, CURLING AND PUCKERING OF LEAVES	NICOTINE PYRETHRIN ROTENONE	GREEN LACEWINGS (CHRYSOPA CARNEA)	FOR A SMALL INFESTATION, REMOVE AND WASH WITH MILD SOAP SOLUTION. WATCH OUT FOR ANTS.
MEALY BUG	1/4" - 1/2" LONG, COVERED WITH WHITE FUZZ WHICH MAKES THEM LOOK LIKE PIECES OF COTTON. THEY HAVE TAILS AND MANY LEGS.	WILL CAUSE NEW LEAVES AND FLOWERS TO DEVELOP POORLY. FOUND ON UNDERSIDES OF LEAF, AT LEAF AXILS & ALONG VEINS.	PYRETHRIN ROTENONE	CRYPTOLAEMUS	REPEAT TOPICAL TREATMENTS EVERY WEEK UNTIL CLEAR. ISOLATE PLANT. WATCH OUT FOR ANTS
SCALE	1/8", WINGLESS, LEGLESS, WITH A WAXY, SCALE-LIKE COVERING. THE LARGER ONES ARE GREY OR BROWN MOTTLED WITH BLACK.	THEY MOVE VERY LITTLE SO OFTEN GO UNDETECTED. FOUND MOSTLY ON FERNS, CACTI, CITRUS AND SUCCULENTS ON STEMS AND UNDERSIDE OF LEAVES. WILL STUNT PLANT GROWTH	PYRETHRIN ROTENONE NICOTINE	APHYTIS MELINUS	ISOLATE PLANT. FOR MINOR INFESTATIONS WASH OFF WITH TOOTHBRUSH WITH MILD SOAP SOLUTION.
ANTS	YES, ANTS ARE A PROBLEM IN GREENHOUSES, PRIMARILY BECAUSE THEY "HERD" APHIDS, MEALY BUGS AND SCALE BY MILKING THEM FOR THE SWEET HONEYDEW THEY SECRETE. IF ANY OF THE ABOVE ARE PRESENT, CHECK ALSO FOR ANTS (ESPECIALLY IN SUMMER AND FALL). IF YOU CAN CONTROL THE ANTS, CHANCES ARE THE OTHERS WON'T BE AS MUCH OF A PROBLEM.				
SPIDER MITE	1/50". VERY SMALL. BODY IS OVAL, YELLOWISH COLOR WITH 2 DARK SPOTS ON BACK (ADULTS)	MITES MAKE COBWEBS ON UNDERSIDES OF LEAVES AND FROM ONE LEAF TO ANOTHER. OFTEN COVERING A NEW SHOOT. CAUSE YELLOW PIN SPOTS ON LEAF	PYRETHRIN ROTENONE	AMBLYSEIUS CALIFORNICUS VERY EFFECTIVE	ISOLATE. CAN BE VERY SERIOUS PEST. ALSO THEY HAVE A HIGH RESISTANCE TO CHEMICAL SPRAYS, MOST ACTIVE IN WARM TEMPERATURES.
SLUGS	1/4" - 1" AND CAN BE MUCH LARGER. SMALL SLUGS ARE DARK GREY OR BLACK. LARGER ONES ARE GREY OR BROWN MOTTLED WITH BLACK.	THEY HIDE IN DARK, DAMP PLACES DURING THE DAY AND COME OUT TO FEED AT NIGHT CHEWING LARGE HOLES IN LEAVES.	SLUG BAITS, POWDER OR PELLETS. THEY HATE TO CRAWL OVER ANYTHING SHARP SO A CIRCLE OF CINDERS, LIME OR SAND AROUND A PLANT HELPS		SLUGS CAN BE A SERIOUS PROBLEM IF LEFT UNCHECKED. BE CAREFUL WITH SLUG BAITS--THEY CAN BE HARMFUL TO CHILDREN OR PETS. IT IS BEST TO PUT BAITS UNDER PIECES OF BOARDS, TIN CANS OR LIDS. SMALL CONTAINERS OF BEER OR GRAPE JUICE WILL ALSO ATTRACT SLUGS--EFFECTIVE THOUGH FOR TWO TO THREE DAYS ONLY.
GNATS	1/10". VERY SMALL, LOOK LIKE SMALL BLACK FLIES.	SCAVENGERS. FEED ON DEAD AND DECAYING MATTER AT SOIL LEVEL. FLY UP WHEN DISTURBED.	PYRETHRIN ROTENONE		NOT A SERIOUS PROBLEM ALTHOUGH LARVAE WILL FEED ON SMALL PLANT ROOTS. IF YOU USE FISH EMULSION FERTILIZER THEY WILL BE MORE PREVALENT.
SOWBUGS OR PILLBUGS	1/4". HARD OVAL SHAPED BODIES. USUALLY GREYISH. WHEN DISTURBED, THEY ROLL UP INTO A BALL, HENCE THE NAME "ROLY-POLY."	SCAVENGERS. FEED ON DEAD AND DECAYING MATTER. CAN HARM YOUNG ROOTS OR TENDER SEEDLINGS.	SLUG BAIT		SAME AS ABOVE. GOOD SANITATION AND VENTILATION HELP.

Figure 100

People and Plants

When children are old enough to contribute their love and labor to the greenhouse, we urge you to let them. The only skill required for planting, watering and harvesting is judgement—and perhaps intuition.

Most children love to work with plants and the plants seem to benefit from the care of children. The results are fairly immediate and highly visible, yet mysterious. How does a seed turn into a plant? It does and it grows quickly. Children have a single-minded kind of energy that can be channeled into extraordinary care for the plants. It's a mutually supportive relationship.

The satisfaction of greenhouse gardening extends to people of all ages. Most of the work is not strenuous and its rewards are great enough to encourage everyone to give it a try. The natural pace of this work is unhurried; the environment is relaxing and absorbing. You can enter the world at a different level in the greenhouse, be transported and totally involved at the same time.

CHAPTER VIII THE STATE OF THE ART

In this chapter we'll give you an idea of what's being done in the field of solar greenhouse design. As you will see, there is no *one* solution to combining greenhouse features. These structures, which are only a representative selection of the kind of work that's going on, offer an amazing variety of approaches. Most of the people featured in this section make their living doing solar design and related work. Many sell detailed plans and offer consultant services (for professional fees) and welcome your inquiry. When writing to them, *always* enclose a self-addressed stamped envelope.

In this edition we have divided the chapter into sections on the following topics:

- Attached solar greenhouses (retrofits)
- Solar greenhomes.
- Independent (freestanding) solar greenhouses.
- Manufacturers.

In some cases they overlap.

Attached Solar Greenhouses

The attached solar greenhouses featured here include a number of styles and approaches that differ from the economical shed roof design seen in earlier chapters. The designs are adapted to specific climatic and site conditions around the country, to cost, and to individual architectural directions.

Figure 101

Jim Burke/Mark Ward

If you are considering building a greenhouse, you should be aware of the availability of used greenhouse material. Over the past four years, Jim and Mark have been dismantling large commercial cypress and glass greenhouses and recycling the material into smaller, more efficient ones.

This material is very durable and attractive. Cypress is a wood ideally suited for greenhouse construction, as it will last practically forever under humid growing conditions, and glass, in addition to having a long lifespan, is attractively clear.

The used material also allows great flexibility in shape and is ideal for custom designs as well as conventional styles. Most important, it is available at a fraction of the cost of new material and uses no additional energy for its creation. Those designing solar greenhouses might find that the lower cost and longer lifespan of these materials outweigh the advantages of more exotic materials.

Mark Ward added-on this (recycled) solar greenhouse designed by Tony Mastrobuono (Fig. 101). Used primarily to supply supplemental heating to the main house, the system relies on a 70 ton rock bed for thermal storage.

The multifunctional greenhouse pictured above (Fig. 102) points out the potential for rooftop greenhouse sites for urban and suburban buildings. Designed by the owner, Robert Patterson, it was built by Mark and Jim for a materials cost of $600. The 276-square-foot structure encloses three sections: greenhouse, sitting room, and dining area.

Recycled materials from a traditional greenhouse were effectively used to construct the energy-efficient attached solar greenhouse seen in Fig. 103. Notice the sturdy old hardware and top venting system (pipe gears activated by chain/wheel pulley) and the steel pipe assemblies used to support the roof. The insulating blanket, shown in its ''up' position, is supported by 2 x 4-inch tracks.

Figure 102

Figure 103

Jim Burke Vermont Recycled Greenhouses
Franklin Rd.
Vernon, Vermont 05354

Mark Ward
396 Cambridge Turnpike
Concord, Massachusetts 01742

Doug and Sara Balcomb
Santa Fe, New Mexico

Figure 104

This unusual and attractive solar greenhouse addition was designed and built by the Balcombs in 1976 for their Los Alamos, New Mexico, home (Fig. 104). The 400 square-foot unit features a sawtooth geometry with clerestory windows. Roof angles were calculated using winter sun altitudes so that the front section does not shade the windows under the rear (north) roof plane. The entire roof is supported by 4 x 12-inch hollow box beams spanning a 20-foot length. A 6-inch drop from west to east allows for drainage. The vertical eastern wall is "stepped" back to echo the sawtooth roof design; only its south-facing planes are glazed for collection. The addition gives the soaring feeling of a Gothic cathedral with the surprise of living pine trees (already on the site) growing up through the roof.

Thermal storage is both passive and active. The passive storage is by direct gain to earth and floor mass within the planting area. The unit also has an active hot air system. Heated air ($95^{o}-100^{o}$ on a clear winter day) is tapped at the highest point in the greenhouse and blown down a 12-inch duct into a rock-filled box below (Fig. 105). Nighttime heating is supplied by the charged-up stones. This system is interesting because it can be added after your greenhouse is constructed. Doug told me, "I could have put the storage in the floor, I suppose, but this way we could use it as a planter box and grow things on top of it." This method appeals to me because it eliminates costly or exhausting excavation work. Plants love bottom heat and the top of the box should be one of the healthiest areas in the greenhouse. Also, should any problems develop in the storage fans, they are above grade, easy to get to and work on.

Dr. Balcomb estimates the finished greenhouse cost about $3000, and "more hours of my labor than I care to calculate." But look at the care and concern that went into the design. What a space! The Balcombs now live in a solar greenhome (Unit I, First Village, p. 119).

Builder, Designer: Doug and Sara Balcomb
Cost: Approx $3,000 (1976)
Size: 400 sq. ft.
Clear Area: Thermopane units
Vents: 19 sq. ft. lower bottom of south wall; 27 sq. ft. upper in second sawtooth. Natural convection or reverse fan.
Heat to home: Natural convection through sliding patio door.
Storage: Above-grade box, approximately 4' wide x 3' deep x 20' long, 240 cu. ft. rock-filled, 1" insulation.

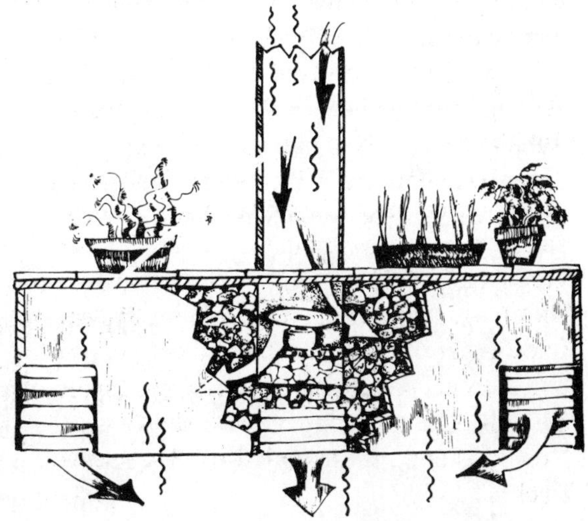

Figure 105

Ced and Betty Currin
Midland, Michigan

Ced Currin is an energy technologist for Dow-Corning and an avid supporter of the solar greenhouse concept.

The Currin solar greenhouse was designed and built for operation in central Michigan winters (known for long periods without direct sun in late autumn and early winter), and to complement the architecture of the house to which it is attached (Fig. 106). This greenhouse was constructed in 1977 and 1978 in Midland, Michigan at 43.7^{o} north latitude. The solar design was by the owner, Ced Currin, and the architectural design by Francis E. Warner, architect of the house. The greenhouse is attached to the west wall of the house off a door from the study.

Figure 106

Following are characteristics and features of this greenhouse:

Size: 12' x 24' 288 sq. ft.

Construction: Frame, 2 x 6 studs, redwood siding, Douglas fir roof beams on 48 inch centers, 3 rows of vertical roof apertures oriented east-west, drywall interior on north wall, painted white; roof overhang, 24 inches; foundation depth, 48 inches.

Glazing: ASG Lo-Iron^{tm} glass, 2 panes, vertical; south wall glazed area, 138 sq. ft.; roof apertures (3 rows), 87 sq. ft.; total south facing glazing, 225 sq. ft.; west wall glazing, 40 sq. ft.; total glazed area, 265 sq. ft.

The glazing is designed with sloping reflective aluminum roof and roof overhang to admit less than 10% of direct sun at solar noon on summer solstice, and between 90 and 95% between each equinox and the winter solstice. The west wall glazing is manually covered with foam insulation during the winter season, and uncovered during other seasons to increase the length of the photoperiod.

Thermal Storage: Nine vertical 55 gallon drums are placed along the north wall and eight horizontal drums are placed behind the south wall glazing. All drums are painted brown and together they contain 7500 pounds (3400 kg) of water.

In addition, 530 cubic feet of water-saturated sand sealed in polyethylene film is used for long term thermal storage. Water heated in the north wall drums is pumped through 100 feet of plastic pipe in the sand to transfer heat into and out of this storage. Heat is also conducted from the sand storage to the wood floor to maintain a warm floor. The water is pumped using a 100-watt concentrating type solar-electric generator only between September and March when direct sun is shining.

Thermal storage capacity:

Drums	7500 BTU/^{o}F	4.0 kWh/^{o}C
Wet sand	16000 BTU/^{o}F	8.4 kWh/^{o}C
Total	23500 BTU/^{o}F	12.4 kWh/^{o}C

For a temperature range of 45 to 80°F (7-27°C):

Drums	0.26 million BTU	80 kWh
Wet sand	0.56 million BTU	170 kWh
Total storage	0.82 million BTU	250 kWh

Vents	
Low vent area, west wall	6.4 sq. ft. (0.60 sq. m)
Area of 3 high vents between roof beams on north wall	8.2 sq. ft. (0.75 sq. m)
Insulation	
North wall and opaque portion of west wall	R-25 (6 inch fiberglass and 1 inch Styrofoam™ brand foam)
Roof	R-14 (2 inch Styrofoam insulation)
Foundation exterior	R-5 (1 inch Styrofoam insulation)
Wet sand storage	R-11 (2 inch Styrofoam insulation)

Construction Costs (Approximate, 1978)

Foundation, wet sand storage, pump		$ 1600
Glazing, installed		1100
Framing, walls, roof beams, drums		3700
Roof, excluding glazing		3100
	Total	$ 9500

Contracted labor	$6000	$20.80/sq. ft.	$225/sq. m
Materials	3500	12.20/sq. ft.	130/sq. m
Total	9500	33.00/sq. ft.	355/sq. m

About 90 percent of the labor was contracted. The oak floor rounds are not included above as they were cut from the owner's trees.

Paul and Judy Vance **Santa Fe, New Mexico**

In planning the orientation of their solar greenhouse addition, the Vances ran into a problem common to the suburban builder. Their property line was a mere five feet from the home's south side. The solution was to build the unit off of the west wall of the home (Fig. 107).

The 180-square-foot front face is constructed with coverings of fiberglass outside / polyethylene inside. Paul designed the clear area in the top of the east side to collect winter sunlight as the sun rises above the roof ridge. This clear section also serves as a high vent, operated manually with a simple pulley arrangement.

When Paul was in my class, I suggested that he also make the west wall entirely clear for maximum collection. The design has worked well; summer afternoon overheating is avoided by providing excellent cross ventilation and exhaust through the high eastern vent. The unit maintains temperatures in the low 40's during sub-zero weather and supplies the home with warm air through sliding glass doors in the shared wall.

Figure 107

Figure 108

Six 55-gallon water drums and an insulated concrete/block north wall (7 feet high by 18 long) provide thermal storage for the all-passive unit.

The Vances and their greenhouse were featured in the educational film *Build Your Own Greenhouse—Solar Style* (see p. 201). In the film Judy says, "I'd like to build a greenhouse and live in it year 'round. I love them." Apparently she meant it, as she and Paul have sold this house to move into a solar greenhome.

Jo Ann and John Hayes — Marlboro, Vermont

Here is another attached unit *intentionally* designed (at the same time as the home) to adjoin the west wall of the home (Fig. 108). The Hayes' greenhouse utilizes south-facing glass only. The western wall is solid/insulated, as that exposure is shaded in the winter. All interior surfaces are paneled with aluminized sheathing material for light reflection to the plants. Air exchange with the home is through a standard-sized door and is totally passive. Because the greenhouse is built above the ground on stilts, interior massive storage (and weight) is reduced. Planting is done in raised beds and on tabletops.

Fran Wessling — Albuquerque, New Mexico

Dr. Francis Wessling was one of the few mechanical engineers in the early 1970s who could see the advantages and the potential of passive solar applications. Fran, wishing to establish a more analytical base to passive design, built and monitored four different systems added to a frame stucco home in

Figure 109

Albuquerque, New Mexico. Three of the systems (the ones in the foreground of Fig. 109) are greenhouses which contain different features. The last unit has a vertical wall with a perforated, decorative cinder block trombe wall directly behind it. (Block like this is available through masonry materials companies and landscape supply firms.) Light passing through this geometric wall has a lovely Arabesque quality and forms pretty patterns at various times of day. A perforated wall of this nature has a much greater surface area for light absorption than a regular massive wall.

The greenhouses all are identical externally and quite different internally. One is a typical add-on structure quite similar to the model recommended in the construction chapter. Another has an active component and uses a fan to charge a rock bed below it. In the third, Dr. Wessling applied an iron oxide mortar to the existing stucco wall of the home. Iron oxide mortar (about 6'' thick) has a thermal conductivity more than twice that of concrete, thus diffusing thermal energy more rapidly. In the photo, notice the simple, adjustable shading system using bamboo mats with conventional rope and pulley controls.

Fran Wessling's technical results are widely published in proceedings of Solar Energy Society conferences and ASHRAE publications. Some of his findings have led directly to practical greenhouse rules of thumb such as the ventilation formula found in Appendix H.

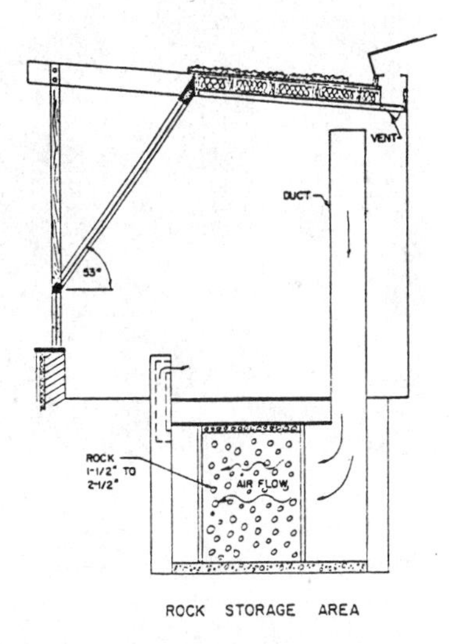

Figure 110

Lynn Nelson **San Francisco, California**

Lynn Nelson has been a leading solar practitioner in the San Francisco Bay area for several years. Her organization, The Habitat Center, is well known for ''hands-on'' workshops and innovative design. As Lynn has had to deal with some of the strictest building codes and ordinances found anywhere in the U.S., she has become expert at adapting solar structures to existing regulations. Her expertise extends to complete home designs and adobe construction (she is currently building an adobe solar greenhouse in the bay area!).

The following retrofit was built during one of Habitat Center's ''hands-on'' workshops (Fig. 111). Lynn describes specific problems and solutions below; they will have wide application throughout the country.

Habitat Center Solar Greenhouse Retrofit: the Daar residence
Design: Lynn Nelson (the Habitat Center)
Construction: Habitat Center ''hands-on'' workshop

In most urban areas, city planning departments establish minimum yard areas. For many city homes on small lots, that means that there's no place to build an attached solar greenhouse without finding yourself in violation of those yard (or setback) requirements. That was the problem with the residence of Sheila and Frank Daar, who wanted an attached solar greenhouse on their home. However, the house has a fair-sized (21' x 7') south-facing porch. So our approach was to create a ''porch enclosure'' that worked as a solar greenhouse.

Since the front of the house had a very distinctive appearance, we decided to leave all that untouched and to fit the glazing in between the posts. We used glass cut to fit for the large areas on either side of the door, and then filled in the smaller spaces between posts and under the glass with greenhouse fiberglass trimmed with redwood lath. And the Daars finally found a use for the oak-and-glass door and leaded glass windows that they'd had in storage for several years: as the door into the greenhouse and as casement windows at the east and west ends.

Photo by Malcolm Collier

Figure 111

Since the amount of glass in the south wall was limited, we tore out about half of the porch ceiling and the roof above it and created a skylight (double-glazed fiberglass) with the joists and ceiling painted white to increase light reflection.

Heat is vented into the first floor of the home by opening the front door and windows, and into the second story bedroom via a vent cut at the top of the greenhouse in the bottom of the bedroom wall. Heat storage in the greenhouse will be provided in water-filled, small, metal containers to start with—and later by the water in a hot tub!

Summertime cooling is provided by opening the door and casement windows to create natural ventilation. The mass is shaded by porch overhang.

Final cost figures aren't all in yet, but the Daars estimate the total cost will stay below $1000.

The most ambitious of our attached solar greenhouse designs (and our passion for "architectural integration") and "hands-on" construction workshops was the two-story greenhouse built on the second and third stories of the A-frame residence of Lorell Long in Penryn, CA (Fig. 112). The 335 sq. ft. greenhouse was built onto the (rebuilt) second story deck (about 24' long x 12' wide) and encloses the third floor deck, which has become a little balcony inside the greenhouse.

Photo by Alan Taylor

Figure 112

To build the greenhouse, we extended the A-frame walls to the south edge of the second floor deck. Those walls will be shingled on the outside to match the existing roofing, and are insulated and sheetrocked on the inside. Thus all the greenhouse glazing is south-facing, except for a 4' x 6' east window. We installed a sliding glass door for entry and used double-glazed sliding glass door replacement seconds for the majority of the vertical south glass and the roof glass, which is set at a 45 degree angle. We then used double-glazed greenhouse fiberglass (Lascolite) for the remaining triangular corner sections, rather than buying glass cut to fit.

The greenhouse solar-heated air is vented into the home simply by opening the sliding glass doors to the living room on the second floor and the bedroom on the third floor. Since there are very few interior dividers between rooms in the house, the heat can easily spread through the entire house.

Solar heat storage is provided by 55-gallon drums lined up against the house glass. The deck (which we had to re-build) was engineered to support the weight of the thermal mass.

Summertime cooling is provided by opening the greenhouse sliding glass door (which serves as a low vent) and then opening the triangular vent at the peak of the greenhouse (which serves as a high vent). The third story balcony inside the greenhouse serves as an overhang that shades the barrels and the living room south glass in summer.

Total cost of the greenhouse and deck was $1950, or about $5.90/square foot of greenhouse floor space.

Figure 113

Rob Jenkins Grand Junction, Colorado

The attached solar greenhouse shown here (Fig. 113) was built for Mesa College, Grand Junction, Colorado, in a workshop held in November 1978. Rob Jenkins designed the structure to employ a *continuous truss system* for framing the greenhouse roof and front face. He feels that it simplifies construction of the two planes and improves thermal performance of the greenhouse. The following is Rob's description of assembling and erecting the truss system.

Assembly: The size of the truss members will depend upon span and design loads. For all but very long-span situations, 2 x 4's will provide adequate strength, and when fully assembled, will be light enough for easy handling. The truss is composed of two members, dadoed at their adjoining faces and bolted with three 2-inch stove bolts (Fig. 114). The angles of roof and south face are not of great importance in the dado process. If the angle of the roof from the vertical and the angle of the south face from the horizontal are the same, the truss members will be identical at the joint. If site conditions and the adjoining building allow, the lengths and end cuts can also be exactly the same.

The radius cut at the assembled truss is best done after bolting, and use of template will facilitate cuts at each truss. The size of the radius may differ. However, it should be noted that as the angles of the south face and roof are varied, the area of the lap of the two truss members will also vary. A more square roof and wall arrangement will involve a smaller lap. Consequently, the radius must be reduced so as not to further diminish the lap area. The Mesa College greenhouse is cut at a 9'' radius with south wall and roof at 80° respectively. In two previous applications, the radius was 12'' and the angles of inclination were approximately 70°.

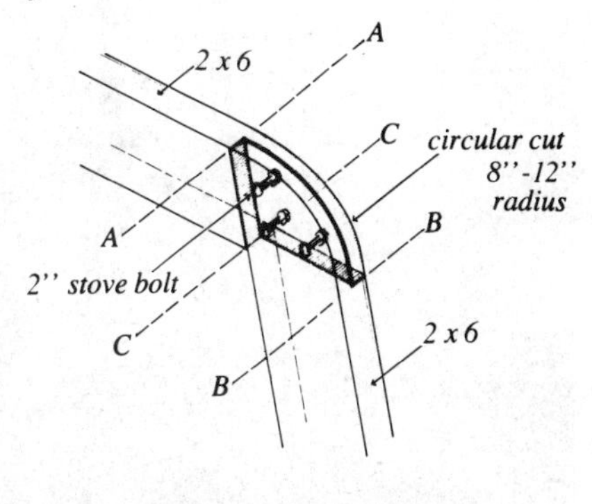

TRUSS CONNECTION

Figure 114

Erection and Glazing: For an attached structure, the upper end of the truss will be anchored to a ledger plate with a joist hanger, and the base will usually be toe-nailed to a sill plate. Because of the continuity of roof and south face, very little intermediate blocking is necessary. Adequate shear strength can be established by an opaque roof section (4x plywood) together with the 5 ft. sections of flat fiberglass sheeting. Trusses should be spaced at no greater interval than 2'-0''. Care should be taken with the length of glazed roof, however, and snow loads should be evaluated.

Most critical in the erection and glazing process is consistency in truss assembly and installation. Because the fiberglass must bend over the knee joint, any difference in planes of the three trusses supporting one fiberglass sheet can mean serious buckling problems. It is, therefore, important initially to insure a level ledger and sill plate. The two end trusses should then be erected and plumbed, after which construction lines should be stretched from points A and B. This will allow all intermediate trusses to be installed level at critical points of ledger plate, transition points A and B, and sill plate.

As the trusses are installed, a temporary tack board can be nailed in at point C to maintain truss spacing and lateral stability until the fiberglass is fully installed.

Fiberglass, 49½'' wide, is extremely beneficial. It provides a full lap at each end truss, and will help accommodate small deviations in truss material and installation.

Following the glazing process, wood lattice battens are installed at each truss and are sealed with clear silicone at the junctions of fiberglass sheets. Batten strips should be kerfed and water soaked where they bend at the truss knee.

Cost: The detail cost for continuous truss framing is less than for conventional eave construction, if two foot spacing is maintained in each case. Fewer materials are used, and if reduced construction time can be achieved, less labor.

Valerie Walsh — Santa Fe, New Mexico

Valerie Walsh is a designer and builder in Santa Fe, New Mexico, who specializes in solar greenhouse additions. Valerie's company, Green Horizon, has been building greenhouses of exceptional

Figure 115

quality and beauty since 1976. Each greenhouse is specially designed to meet the homeowner's needs. Valerie's solar additions provide more than just a thermally efficient growing space; they become the most exciting living area of the home. Greenhouse owners find this the ideal environment for relaxing, dining, and entertaining.

Figure 116

This addition (Fig. 115) in Santa Fe, engineered by Noel and Jody Norskog, shows Green Horizon's high level of craftsmanship. The mahogany frame construction consists of laminations of fourteen individual curved modules. The glazings for each module include three sections: an overhead curved acrylite panel, and two vertical panels of insulating glass, the bottom of which opens for ventilation. In summer, air is circulated through the upper roof vents.

Thermal mass has been designed as a functional element of the structure. For example, the sculptural walls of the plant beds are made of concrete-filled block that are stuccoed for sitting. The north wall of the greenhouse is the existing adobe wall of the adjoining home, an excellent heat sink. Additional thermal mass is provided by a concrete floor slab covered with decorative clay tile. The masonry mass is supplemented by 300 gallons of water in the hot tub. Aztec Radiant Heat Panels[R] fit unobtrusively into the ceiling as a back-up heat source. If supplemental heat is necessary, low-temperature radiant heat proves to be 30% more efficient than hot-air systems.

The owners enjoy the functional planting areas. Ground beds are filled with vegetables and the high ceiling allows for tall and hanging plants. The work area has a deep sink, potting bench, and storage cabinets.

The greenhouse is located off the kitchen and living room; supplemental heat is transferred into the rooms through a connecting window and door. For dining and entertaining, there is plenty of space for a table and chairs. The family also enjoys relaxing in their hot tub surrounded by greenery and a spectacular view.

Green Horizon continues to custom-build beautiful energy-efficient greenhouses. The company is currently expanding module construction and developing nighttime insulation systems.

Ozark Institute — Eureka Springs, Arkansas

This Berryville, Arkansas solar greenhouse is an outstanding example of the 'spin-off' the U.S.D.A. Solar Outreach program achieved. The Ozark Institute, headed by Edd Jeffords, sent Bill O'Neill to a Little Rock workshop (sponsored by the U.S.D.A.) to get technical training and practical experience. Mr. O'Neill went back to Eureka springs as the solar project director and, with Albert Skiles, designed and built the Berryville greenhouse and two others. The Institute had received a $1,000 grant for one greenhouse. By recycling materials and careful purchasing, they built three greenhouses for that amount.

From Mr. Skiles report for the Office of Human Concern in Arkansas:

"Two such greenhouses were recently built onto the homes of J.D. and Dorothy Trahan of Madison County and John and Coleen Strandquist of Carroll County. Winter heat gain will be more efficient than conventional designs because of a reflective-insulative shutter along the vertical south wall. Open while the sun is shining, the shutter will admit about 30 percent more solar energy onto storage containers within the structure. At night the shutter is closed to greatly reduce heat loss.

Figure 117

"The glass panes on the south wall are removable to control heat during hot, humid periods. Prevailing south breezes can then enter through a continuous fixed screen. This gives the greenhouse a generous vent to floor ratio of 1:2. The glass panels can easily be replaced the next fall in an airtight manner.

"The Trahan greenhouse has a thermostatically controlled blower near the ceiling which will deliver heat (on at 110°—off at 90°) to the house whenever it becomes available. By attaching a meter in the circuit, we can accurately determine how much surplus heat is produced by the greenhouse.

"The Standquist greenhouse will be connected directly to the living space with no dividing wall. Heat gain to the house will be found by comparing the fuel cost per degree-day ratio with those of previous winters."

Solar Sustenance Team Office **Rt. 1 Box 107AA, Santa Fe, N.M.**

The Solar Sustenance Team office was added to an existing two-room adobe in Nambe, New Mexico. The design problem was one familiar to many homeowners: the corner of the existing building faced due south. Does one build on the 45° east- or 45° west-facing wall? I felt that building the solar greenhouse *around* the corner would give us a good opportunity to work with a rather unusual geometry (it should have benefits and drawbacks not found in a single-axis unit).

The outcome, to date, has been satisfactory. First, the greenhouse receives much more light than a south-wall attached model. Because the major collecting planes are 45° to the east and west of south, spring, summer, and fall sun is in the greenhouse all the long days through. This is a real positive factor for plant growth.

I thought that the off-south glazings would result in an extremely cold greenhouse in mid-winter. This was not the case. The greenhouse maintained minimum temperatures above 48° in −12° weather (after 4 or 5 heavily cloudy days). This is primarily because of the tremendous amount of thermal mass found in the shared adobe wall between house and greenhouse.

The total amount of mass in the greenhouse works out to be the equivalent of over 9 gallons of water per square foot of glazing. Hence, it is very stable in temperature both summer and winter.

Figure 118

I also expected the large expanse of western-facing glazing to contribute to horrendous summer overheating problems. So, I designed-in large low vents with boathatch exhausts in the roof. In addition, the pyramid-shaped observatory was designed to be a thermal exhaust chimney for the house as well as the greenhouse. Four of the triangular Lexan-covered panels open all the way out for stargazing. One of these panels has a summer exhaust position when it's partially open. In this mode, the observatory heats up to about 115°F and pulls air through the house and greenhouse at the rate of about 2000 cfm. Quite an effective passive cooling system.

Other features of the office are a composting privy, a gray-water recycling system, photovoltaic fan, and the batch water heater seen above the greenhouse in the corner of the house roof.

All of the details in the greenhouse could make an entire chapter of a book. However, there are some subtle and important aspects of the office that I'll mention here. I think of the structure as a blend of the old and new. Mud-plastered walls and flagstone floors are combined with photovoltaic fans, heat

Figure 119

motors, and a solar greenhouse heating system. With the exception of the glazings, the materials used in building are low-energy and natural products. For example, we experienced a severe concrete shortage shortly after beginning. So we went back to the old ways . . .river rock foundations and mud plaster. There's an amazing amount of craftsmanship in the woodworking and details of the office. Native woods, lovingly assembled by McChesney Hortenstein, have been used in the cabinets and doors. Privacy walls, constructed after a careful examination of prevailing wind patterns, do double duty as protective wind barriers in winter and air-flow funnels in summer.

The old and new are blended into a space that's aesthetically pleasing (for me) and energy efficient (for all).

Designer: Bill Yanda
Glazing: 310 sq. ft.
Greenhouse area: 350 sq. ft.
Total office area: 1106 sq. ft.
Solar Heating Fraction: 65%
Total cost (includes office remodeling): $18 sq. ft. (1979)
Builders: McChesney Hortenstein, James Bunker, Andrew McGruer, Bruce Currin.

Figure 120

Tim Michels

Londe, Parker, & Michels, Consultants
7438 Forsyth, St. Louis, Mo.

Figure 121

Many of the brick homes found throughout urban America are ideal for a solar greenhouse addition. They will generally have either solid brick or brick with a clay-like infill. Tim Michels and his associates completely retrofitted this home in St. Louis. They began, as anyone serious about saving energy would, by adding insulation to the roof. After that they applied rigid insulation to all exterior brick walls with a stucco overcoating. Then they began the solar part. A small second story window was enlarged and double glazed for direct gain and Trombe wall combination. The solar greenhouse below is a seasonal solar heater quickly erected in the fall and removed in the spring. In this manner the south facing brick wall is converted from a heat losing surface to a low temperature radiant heater in the winter time.

Michels used computer runs to determine the economic payback of the total conversion. The total job, insulation, interior remodeling, solar additions, came to slightly less than $4,000. Measured against any source of conventional heating fuel the investment will be cost effective.

A Portable Homemade Greenhouse

So . . .you're a renter, or you move a lot. You want a greenhouse but don't want to leave it behind for the next occupants. Here's an alternative. Build a portable, but substantial, greenhouse that you can pack up and move to your next home.

Figure 122

This small lean-to was designed as a demonstration model to be moved around to fairs, energy exhibits, schools and the like. It's lightweight and can be assembled by two people. The largest panel is 8 ft. x 8 ft. and the panels fit into a rack on a pickup truck. Even though this is a display model, the greenhouse is fully functional, containing all the criteria for successful operation found in Chapter IV (insulated opaque walls, double skin, shaded roof, etc.). When we set it up, we put in black water barrels and plants.

The five planes (you don't need a north wall, that's for display) are held together by slip pins in regular hinges. The intersections of the edges are sealed with foam insulating tape. The whole unit sets up in about 10 minutes with two people. Just set it down in front of your south window and you're in business. In a home application, it could be mounted on railroad ties (as explained in Chapter VI) right on the ground, or on a low block wall.

Designers: Paul Bunker, John Galt, Bill Yanda
Size: 64 square feet.
Builders: Same as designers with help from the New Mexico Organic Growers Assn.
Clear Area: 125 square feet
Hardware: 2'' sliding bolt hinges hold planes together; framing members pre-drilled then screwed together with No. 6 2½'' wood screws

Working blueprints of this greenhouse are available for $10.00 from The Solar Sustenance Team, Rt. 1 Box 107AA, Santa Fe, N.M. 87501.

Solar Water in the Solar Greenhouse

New uses for solar greenhouses are popping up every day. One of the most applicable is a solar hot water system *inside* the greenhouse. When this is done, you have, in effect, moved the location of the solar water heater about 15-20° south in latitude, compared to a similar heater exposed on the outside of the roof.

Cost: $256 (includes double fiberglass walls, north wall and No. 1 clear fir 2x2s for strength and uniformity). Funded by the New Mexico Energy Resources Board.

A $100 ''Breadbox'' Pre-Heater in a Solar Greenhouse Roof

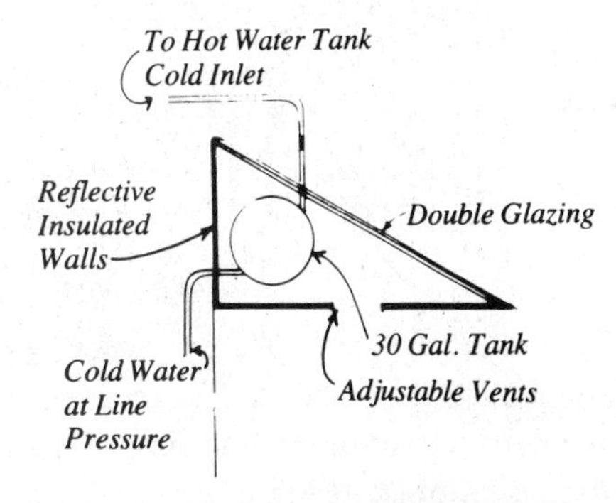

Greenhouse Roof Must Support Load (Approx. 300 lbs.) Spread on 4 Rafters, Beef-up Rafters (Support from Below, or Use Diagonal Braces)

Figure 123

Some people are placing typical flat plate water heaters into the front sloping panels of the greenhouse and letting the hot water naturally thermosyphon up to a tank in the apex. A very simple "batch" or "breadbox" (a term coined by Zomeworks) heater can be built into the apex of an attached solar greenhouse in this manner (Fig. 123).

It is my experience with such a heater in the Solar Sustenance Team office that it will provide the majority of the hot water needs for a couple from April through September. Throughout the remainder of the year it functions as a pre-heater to the conventional hot water tank. It can be built by the homeowner for under $100 or contracted for under $250. In the Sunbelt area, it might provide 75% of the hot water for a conservation minded family of four throughout the year. Batch heaters similar to this were built by the thousands in Southern California and Florida in the '30's and '40's. Many are still working today turning out free hot water on a regular basis.

Don't let anybody fool you. Solar hot water definitely feels better than the conventional variety. You just dance around under the showerhead laughing at everybody who is paying money to have their water heated. If you store the graywater, like we do at the Solar Sustenance Team office, you get quite spoiled on 30 minute showers.

The following article by Rich Schwolsky explains the rationale and plumbing specifics of a more sophisticated unit in greater detail. It is reprinted with permission of a great magazine, *Solar Age*.

Preheating Water in Greenhouse

by Rick Schwolsky
Courtesy: *Solar Age* Mag.
June, 1979

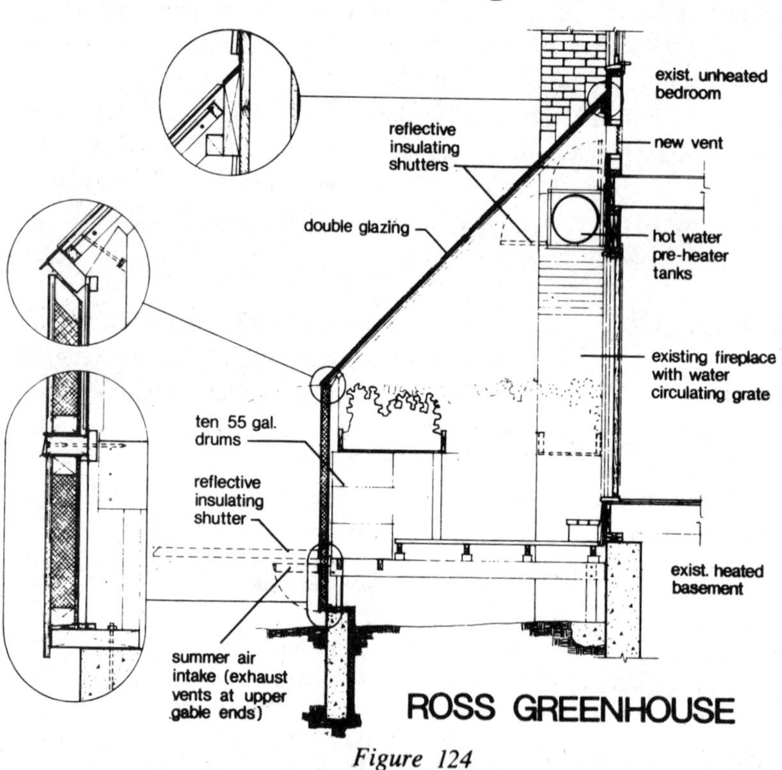

Figure 124

If a greenhouse has been designed to create a non-freezing environment for plants, an opportunity exists to include a simple and inexpensive water heater.

In a greenhouse built by Alan Ross and the Brattleboro Design Group in Vermont, domestic water preheat was a very important consideration (Fig. 124). Used, glass-lined, gas-fired water-heating tanks (30 to 40 gallons) were stripped of their outer jackets and insulation, flushed, pressure tested, and painted flat black. An enclosure was constructed for each tank, using urethane sheathing and plywood (Fig. 125). The enclosures isolate the tanks from the greenhouse space and increase the effective aperture of each tank. The enclosures incorporate doors with reflective surfaces, which can be focused manually in winter, summer, and spring-fall seasons. The doors, closed manually, insulate the tanks from night-time radiation losses through the greenhouse glazing.

Tanks, enclosures and some pre-piping were hung horizontally from the ledger in the peak of the greenhouse. The flue openings in each tank were used as the support detail by placing 1½-inch galvanized pipe through the flue so that it extended beyond both ends of each enclosure. Chains hung from hooks in the ledger were connected to hooks fastened to the ends of the galvanized pipe, and the units were placed on either side of the existing chimney (Fig. 126).

The connection to the water system is very simple. Using ¾-inch copper tubing, cold water supply entering directly from a well passes through the two solar tanks in a series to the existing oil-fired boiler. A three valve bypass was installed near the boiler to allow the removal of either or both solar tanks, which were connected with unions, without interrupting hot water service to the house. Each tank was provided with a temperature and pressure relief valve for safe operation.

Figure 125

Adapting this breadbox-type water heater for use in the greenhouse has advantages apart from economics and reliability. The space in the peak is usually unused and contributes to heat loss. A water heater puts this space into productive use and isolates it with an insulated surface from the rest of the greenhouse. The enclosures also create an overhang that will substantially contribute to summer shading.

Figure 126

The system operates on demand—the collector-storage tanks directly preheat domestic water. There are no differential controllers, no periodic changing of fluids, no motorized valves, no pumps, and no heat exchangers. The system can be drained completely to prevent damage which would otherwise be caused if the greenhouse temperature is allowed to fall below freezing.

The requirement to manually control the insulating doors and to fill and drain the system is not much to ask of owners already committed to maintaining a greenhouse. Designers are not overburdened to integrate solar domestic-water preheat, if they are committed to combining the many concepts that shape the *solar* greenhouse.

Rick Schwolsky is a partner in Sunrise Solar Services, Suffield, Conn., and Brattleboro, Vt., and chairman of the Nat'l. Assn. of Solar Contractors, Suite 928, 910 17th St., N.W., Washington, D.C. 20006.

Solar Greenhomes

The term Solar Greenhome was first brought to my attention by Jack Park of Helion. I feel it aptly describes a house which uses a solar greenhouse as its *primary* heating source. The integral house/greenhouse relationship has many aesthetic and solar advantages over an add-on unit, but is usually quite sophisticated in design and carries a new home construction price tag.

Solar greenhomes are based on what Dr. Balcomb has described as a 'Two-zone' approach to solar heating. The greenhouse, by nature of the plant occupants, can tolerate much greater temperature swings (40°F–90°F) than the humans in the home (65°F–75°F). These differing temperature criteria are the basis of solar greenhome design.

By Cedric Green

Figure 127

Although the solar greenhomes shown in this section are found in drastically different climate zones and use varied materials and building techniques, they all share some similar attributes that aren't usually found in add-on greenhouses.

• EXCELLENT ORIENTATION and communication with the home. When you start from scratch it's no problem to orient the house for maximum winter sunlight and to place doors, windows and vents exactly where they're needed.

• THERMAL MASS within the home to absorb surplus greenhouse heat. Adobe homes found in the Southwest and older brick homes have this feature, but the vast majority of structures in the U.S. do not. Most of the homes in this section are hybrid systems; they have both passive heat storage built into the structure and active warm air circulation to thermal storage below the floors of the living space.

• HIGH INSULATION with minimal loss through north, east and west walls and windows. The designers and builders here greatly exceed the minimum insulation standards for walls and roof. They know it's easier and less expensive to install large amounts of insulation initially than it is to build a bigger collector and storage system.

• PLANNED SPACIAL AND TRAFFIC PATTERNS designed *around* the greenhouse. As the greenhouse is the core of the home's life functions, it is possible to attain a degree of architectural integration and functionality that is rare in add-on units. Often, the greenhouse is used as the main entrance for the house. It becomes an energy saving air lock for the structure. This feature alone can save 10-15% on winter heating bills for a family with kids and pets.

• THERMAL ISOLATION during warm periods. The greenhome designer must insure that the greenhouse component doesn't overheat the home in the summer. This is usually not a problem with the add-on unit; it's much more of a consideration with the totally integrated greenhome. In climates which have fairly cool, dry summers (the Rockies for example), there will be adequate nighttime low temperatures to allow the thermal mass in the structure to cool off. Many other parts of the U.S. have an adequate

Figure 128

summer diurnal swing (nighttime temperatures regularly below 70°F) to allow the same thing. In these areas summer cooling is accomplished by built-in shading and/or passive ventilation. In warmer and wetter climates (like the Southeastern U.S.), the solar greenhomes will demand active exhaust systems, removable glazing panels, or total greenhouse isolation from the home in summer. The Parallax homes, for instance, can completely shut off the greenhouse from the living spaces.

The solar greenhome holds an exciting promise for future generations. Beautiful and functional homes with practically no demand on conventional energy sources for heat and fresh vegetables. In almost all of these homes you are looking at construction costs *at or below* a conventional energy consuming house.

UNIT I FIRST VILLAGE

This is a great place to start the solar greenhome section. I was going to write a glowing section in praise of this home and the designers, Wayne and Susan Nichols, but I solicited and received the following text by Doug and Sara Balcomb. Here's a perfect opportunity for you to get first hand information on what it's like to live in a solar greenhome.

Without a doubt this house is the best performing, well monitored solar structure in the world. The 93% solar fraction is obtained without any sacrifice in comfort. About 73% of the solar heating is totally passive through the massive shared wall between home and greenhouse and the naturally circulating warm air. The other 20% solar contribution is from the rockbeds below the floor which are charged by fans tapping the warm air at the top of the greenhouse. The remaining 7% comes from electric baseboard heat. Even though Santa Fe is an excellent climate for solar applications (cold winters but lots of sunshine) it's important to note that this design, as it is, would provide over 50% solar heating anywhere in the United States.

Now I'll let the Balcombs do the talking.

The Solar Greenhouse Experience by Doug and Sara Balcomb

Living in a combined solar greenhouse-home has been a rewarding experience for us in terms of economy, the ability to grow plants year-around, and the delightful ambience it provides our home. The house is Unit 1 of First Village designed and built by Susan and Wayne Nichols, developers of the First Village Solar Community and with Bill Lumpkins, the well known Southwestern architect.

Our house is two stories with a greenhouse enclosing the south-facing triangular space formed by the L shape of the house. The wall between the house and greenhouse is a massive 14-inch thick adobe wall. The remainder of the house is well-insulated frame construction using 2 x 8s with 8-inch batt for the walls, double glazed windows and four inches of foam insulation in the flat roof. The house sits on a gradual south sloping lot and is dug into the hillside to a depth of four feet on the north side. The below grade portion of the wall is filled concrete block, waterproofed and then faced with rigid insulation on the outside.

Thermally, the house works very well. We have baseboard electric heaters for our back-up and also a small woodstove and fireplace. Santa Fe has a cold winter climate (approximately 6000 degree days), and the heating season lasts from early October to mid-May. Below 0°F weather is common in December and January.

In the winter of '77-'78 we set the thermostats at 65° and left them alone. We used about ½ cord of wood throughout the winter, primarily for the beauty of the fire, not for heating. Our baseboard electric use, which is metered separately from the other household electricity, was 857 kilowatt-hours, costing $38 at our present 4.5¢/kilowatt-hour rate. This was our entire heating bill for the year.

The home tends to remain quite constant in temperature, generally between 71°F and 75°F in the summer and between 65°F and 71°F in the winter. The daily temperature swing in the house is typically 4° in the summer and 6° in the winter. (Note: This is less than half the temperature swing in a conventional home.) Because of the warm wall and warm floor we find that we are quite comfortable at 65°F air temperature—more comfortable than we had been previously in a frame, forced-air heated home at 70°–72°F. There is an almost total absence of the drafts and hot spots that are common in a forced air home.

After about three days of stormy weather, the house drifts down to 65°F and some backup heat is needed until it clears again. Even during a storm, the greenhouse collects and stores sufficient heat so that the electric backup is never used during the day. It comes on only in the middle of the night during off-peak periods.

The greenhouse environment is quite different than the house. It runs cooler in winter and warmer in summer. It normally experiences a 30–35°F temperature swing during a sunny winter or summer day. In winter, the temperature range is between 52°-85°F with 46°F being the all time low. It has no auxiliary heater and the 46°F record was after a long storm with outdoor temperatures in the −15°F range. In the summer the greenhouse typically varies from 65°–90°F on a sunny day. Daytime summer temperatures can be hot in Santa Fe, but the nights are very cool, dropping to the high 50s and low 60s.

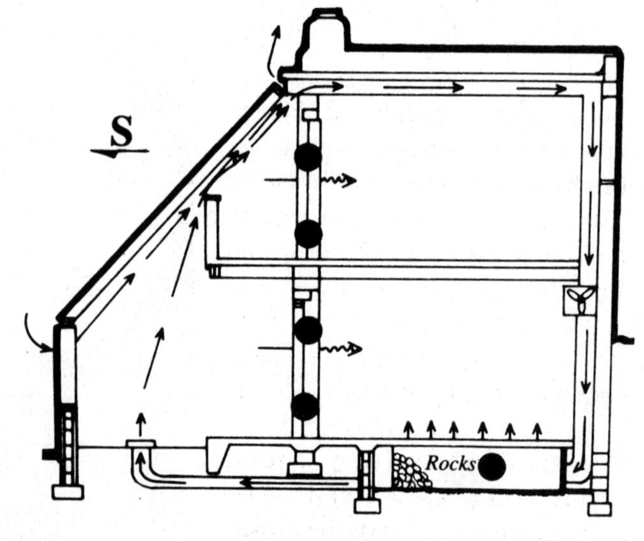

Figure 129

Why does it work so well and will it work in other climates? The house has been monitored by Los Alamos Scientific Laboratory, and we can give partial answers to these questions. The predominant solar heating effect is direct sun entering the greenhouse being absorbed by the floor and adobe north walls. The heat absorbed by the floor (flagstone on dirt, and planting beds) maintains the greenhouse at night so that heat loss by the house to the greenhouse is minimal. Heat absorbed by the massive wall between the greenhouse and the house conducts slowly through the wall giving 85°F temperature peaks on the inside (living space) surface during the evening hours. Thus, this wall becomes a huge low temperature heater along the entire south side of the home.

A secondary source of solar heating the house is through an active hot air loop. During a winter day—typically from 10AM–4PM—the greenhouse is a quite warm 85°F. The warm air is drawn off the top of the greenhouse through two air ducts by a pair of 1/3 horsepower fans and forced through two 25-ton rockbeds beneath the floor of the home. The air is returned to the greenhouse at about 70°F through grills in the floor, completing the loop. The rockbeds are large; 30'' deep pancakes immediately beneath and supporting the concrete-slab floor. The floor temperatures are not especially warm—usually 65°–70°F—but feel nice to the bare feet and are substantially warmer than surrounding non-heated floor areas. There is no problem in partially covering the floor with throw rugs or any other carpeting, and, in fact, part of our floor is hardwood laid over the slab. Distribution of heat from the rockbed to the house is totally passive—by conduction up through the floor.

The advantage of the air loop-rockbed combination is two fold. First, the greenhouse is kept about 10°F cooler during sunny winter mid-days; it has been observed to rise to 95°F if the fans are turned off. Second, about 15,000 BTU per hour are removed from the greenhouse (which would be largely lost) and placed in thermal storage beneath the feet of the occupants. The two fans draw a total of about 500 watts and cost approximately 13¢ a day to operate. They are controlled by a differential thermostat that turns them on whenever the greenhouse air is 15°F warmer than the rockbed. The fans are off when the greenhouse air is 5° F (or less) warmer than the rocks.

An advantage that a greenhouse has over an active solar collector is increased collector efficiency. The system is always 'on', so that even on cloudy days there is some net gain. For example, we have observed that our greenhouse invariably warms up during a winter day, even if it is snowing outside. Air temperatures will increase from a typical 52°F nighttime low to at least 65°F in the worst weather. The floor and walls warm up even more, so that the greenhouse becomes self-sufficient even during stormy weather. This increased collector efficiency compensates for greater nighttime losses of the greenhouse, compared to those of an active collector.

Although the solar heating (and resulting lower utility bills) is of major importance, it is only one aspect of the benefit to human comfort and well-being that an attached solar greenhouse provides. The *quality* of the heat *is as important* as the *quantity*. There are no drafts, no hot or cold spots, no blast of hot dry furnace air, no filters to change. It is completely reliable, working right through power outages and fuel shortages.

Figure 130

It is important to emphasize the quality of life that a well designed solar greenhouse affords the owner. Imagine a blustery, snowy winter day . . .cold and raw. Then imagine what it is like to step directly into a room where roses and petunias are in full bloom, where lemons are ripening on the tree, where the smell of fresh flowers and green plants pervades. Imagine what it is like to sit in a space like that and sunbathe when it is 20°F outside. Finally, and best of all, imagine picking all the fresh vegetables you and your family need for your winter dinners, never having to worry about the price (or the age) of a head of lettuce!

This 'impossible dream' is a reality for us, one we hope every family in America will share.

House: 2 stories, 1,950 sq. ft. in house, 350 sq. ft. in greenhouse.
Greenhouse glazing: 400 sq. ft., lower-vertical, upper 60°.
Rock beds: 2 under the ground floors, 2 ft. deep x 10 ft. wide x 19 ft. long; 24 yds. of 4-6" rock.

The Balcombs, along with the designers Susan and Wayne Nichols, and the author of Passive Solar Energy Book, *Ed Mazria, make up Passive Solar Associates. They run excellent technical design seminars all over the country. For information write: Passive Solar Associates, P.O. Box 6023, Santa Fe, New Mexico 87501.*

Figure 131

Parallax Corporation, Hinesburg, Vermont

The design team of the Parallax Corporation has become a national leader in solar greenhomes and commercial structures built at or below the cost of conventional energy inefficient buildings. Most of their work is in the cloudy and cold region of upper New England. Here, they have to deal with climates in excess of 8000 degree days and storms that can block the direct sunlight for 5-7 days in January. The greenhouse is a natural choice for such climates because it makes the most of diffuse weather conditions.

The solar greenhouse (Fig. 131) is built below the living space of the structure. The other solar component in the home is the direct gain living room above the greenhouse. The living room features a massive core fireplace that absorbs the heat streaming through the south windows. Another feature is a heat recovery system. This is simply a fan linked to a differential thermostat that taps warm air at the ceiling level and blows it down through the rock storage whenever the thermostat senses that the ceiling air is warmer than the thermal storage. A good example might be a cold January night when the occupants were enjoying a nice toasty fire. In a conventional home very hot air would stratify at the ceiling, and increased heat loss would occur. A Parallax home would use this "nighttime heat gain" to charge the thermal storage. Ingenious and inexpensive heat saving. The same mode is operant on a clear winter day.

The greenhouses can either charge the thermal storage or be linked directly to the living space. Again, this is controlled by thermostats that direct the heat where it is most needed. Notice the built-in shading for the greenhouse glazing in the house. Because the summers in upstate Vermont can get very hot and sticky, this structure is designed so the greenhouse is thermally isolated from the living spaces during those periods.

Parallax also uses various glazings to meet specific needs. These homes have Acrylite SDP acrylic glazing in the greenhouse and double glass panels in the living spaces. This gives the owners a nice view out and lets the plants and the thermal mass of the greenhouse enjoy the benefits of diffuse radiation.

One of the most impressive aspects of Parallax structures is their building cost. Most people still believe that solar homes must be appreciably more expensive than conventional structures. The home shown here was built below the cost of regular housing at the time of its construction (1977-78) by licensed contractors with no previous solar building experience. Here are some specifics:

The greenhouse is 36 feet long with glazing 8 feet high. The rockbed behind the greenhouse contains 1000 cu. ft. of rocks. The 2,300 sq. ft. house was built for $56,000 and is 65% solar heated.

Doug Taff and Bob Holdridge, principals in Parallax, have written some excellent technical papers on passive and hybrid solar heating and solar use above the 48° latitude zone. See the Passive Conference Technical Publications.

David Wright
The Sea Group
P.O. Box 49,
Sea Ranch, CA 95497

Figure 132

David Wright continues to be a leader in passive solar design. Since the first edition of this book, David has been busy. First he moved to Sea Ranch, California—a radical change from New Mexico. Then he came out with what I believe is the most readable book in the solar field: *Natural Solar Architecture*. Now he heads up Sea Group, a design team doing passive structures all over the world.

Figure 133

Most of the solar homes he has done are 'One-Zone' structures. However, they are so tasteful and sensitive to local microclimates and ecological criteria that they must be included. I've been in many of Wright's homes, and the owners usually designate areas toward the south glazings as greenrooms and plant growing space.

The Clark Kimbell and Charlotte Stone home in Santa Fe (Fig. 132) includes a "green room" or plant area. The entire room, however, acts as a solar greenhouse, it is designed to be a passive solar collector. The two-story south face is double glazed with commercial storm doors. The continuous east/north/west wall is adobe, sheathed with styrofoam and stucco. The north corners of the house are rounded. Wright adopted this feature from a "D" shaped New Mexico pueblo (Pueblo Bonito). The configuration cuts down on north wind resistance, thus reducing heat loss.

Figure 134

Figure 135

Along with Wright's careful calculation and use of summer/winter sun angles, insulation and tightness are the most important solar features of this home. The roof and sub-floor areas are heavily insulated. So is the front face, with an ingenious canvas-covered styrofoam shutter system. Wright has aptly dubbed this home the "adobe Thermos bottle." It is over 80 percent solar, with one wood stove supplying backup heat.

It is also a delightful living environment, offering all the creature comforts within an ecologically sound design.

Figure 133 shows another Wright designed home. This totally passive dwelling is owned by a Santa Fe builder, Karen Terry. It has an unusual orientation for a solar building, a north-south axis. The home has been designed and built so tastefully that you really have to be looking for it to find it.

One of Wright's newer homes, Sundown, is at Sea Ranch. The main living area is to the lower right in the photo. This part of the house is sunken below grade and has sod and ice plant on the roof. It is a direct gain structure, and the glass faces south. The above ground room on the left faces the Pacific, is a sleeping loft and ideal for observing the whales that swim by the window.

Wright had the insulating barrier assembled by a sailmaker. It fits snuggly on tracks inside of all the large glass surfaces in the house. If you look carefully at the home photo(Fig. 134), you can see these devices in various daytime shading positions. Figure 135 shows the insulating blanket.

Sundesigns, Gregory Franta AIA, Glenwood Springs, Colorado

Greg Franta has been actively involved in the solar greenhouse field since its beginnings. He was a founder of the Roaring Forks Research Center in Aspen and is now working with the Solar Energy Research Institute in Golden, Colorado. The following text describes one project he was involved in privately.

The Smith-Hite Studio (Fig. 136), one of hundreds of solar greenrooms in the Rocky Mountains, is located near Aspen, Colorado, at an elevation of approximately 7500 feet. The architect is Gregory Franta, AIA, of Sundesigns; the owners are Debbie Smith and Henry Hite; and the builders are Richard Farizel and Chuck Ravetta.

Figure 136

The design of the Smith–Hite studio integrates a 600 square–foot weaving studio with a 224 square foot greenroom. The greenroom provides both solar heat collection and space for food production. All of the exterior walls have a thermal buffer between the ambient air and the interior living/working space (Fig. 137). In general, the primary design parameter of the studio greenroom was energy and resource conservation in addition to the space functions.

The greenroom has 312 square feet of south-facing glazing (15° E. of S.). Solar heat is passively absorbed and stored in 3300 pounds of contained water and over 50,000 pounds of gravel and concrete. The water storage is contained in thirty gallon plastic drums located along the north, east and west walls. The primary portion of the remaining passive storage is located in the floor of the greenroom; a 4 inch concrete slab tops 18 inches of gravel. Two inches of styrofoam insulation is located below the gravel storage to reduce heat loss into the ground.

A system of moveable insulating louvres, "Sun-louvres", is incorporated into the roof glazing. The Sun-louvres provide movable insulation at night, act as heat boosters for the hot-air collection system and provide solar shades when desirable.The Sun-louvres have a black heat-absorbing surface on the top side, a reflective surface on the bottom, and insulation in between. In the open position, solar energy is allowed to penetrate into space as much as desired. At the same time, solar energy is absorbed on the black surface of the louvres to boost the air temperature in the higher portion of the space. The hot air is pulled off the top of the greenroom by a single fan. The air is transported through ducts to thermal storage under the floor of the studio. The air returns to the greenroom after it loses much of its heat to the rocks, cooling the greenroom. At night, the louvres are in a closed position to reduce heat loss. Movable insulating panels slide up to insulate the vertical glazing. The passive distribution system allows the natural flow of heat from either the storage bed or the greenroom into the studio.

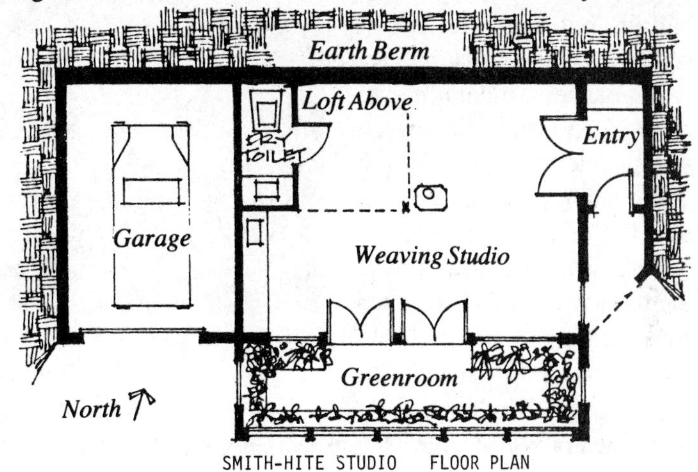

SMITH-HITE STUDIO FLOOR PLAN

Figure 137

The greenroom not only supplies heat to the studio, but also acts as a thermal buffer and reduces the heat loss. The glass wall with French doors located between the greenroom and studio allows the greenroom temperatures to fluctuate with little effect on the studio temperatures. This results in the greenroom buffering the south wall of the studio from the ambient weather conditions. The north, east and west walls of the studio also have thermal buffers: an earth-berm covers most of the north wall; an air-lock entry and mudroom are located on the east wall; and, a garage is attached to the west wall.

Other passive and active heating components are incorporated into the studio. Skylights with movable insulating panels provide direct solar gain onto the thermal mass of the floor and north wall. A heat recovery system transports the hot air from the top of the studio into storage. The only auxiliary heat source is a wood burning stove.

In addition to the energy conservation measures, other resource conservation applications have been designed into the studio. A dry organic waste treatment system, "Humus Toilet", is used to conserve water. The other plumbing fixtures are water-conserving fixtures. Recycled doors and windows have been incorporated into the design. Indigenous and non-energy intensive building materials have been used whenever applicable. The Smith-Hite case study is one of many individual solar greenrooms having impact in the Rocky Mountains.

Clemson University-Rural Housing Research Unit, Clemson, South Carolina 29631

The Rural Housing Research Unit at Clemson has been one of the few universities in the U.S. seriously examining solar greenhouse design and performance. Their approach to building greenhomes incorporates more active collection and control features than most in this book. In a paper, "Construction and Performance of a Solar Greenhouse-Residence," M. Davie, et al., give some realistic examples of problems encountered when the solar home designer meets the builders at the site:

Figure 138

"The coordination between the various trades offers a good insight into the various inherent construction problems. The activities associated with the rock storage, for example, not only concerned itself with the placement of the ballast, but also with the installation of insulation and vapor barriers, monitoring apparatus, and supply and return ducts. Due to the difficulty in getting dump trucks close to the building enclosure, the ballast was dropped outside the foundation wall. It was subsequently shoveled by hand into its position in the bin at considerable expense to the contractor. In addition, the placement of ductwork after the ballast was positioned provided a further expense since the rock had to be removed to allow the installation of a 1' x 2' (300mm x 600mm) distribution plenum.

"The original prototype called for a manufactured lean-to glass greenhouse. To comply with a request by the Department of Horticulture, it was necessary to order, erect and have services installed to allow for August planting. The greenhouse manufacturer implied the relative ease of erection of its greenhouse units—this proved not so. Complications pertaining to the erection procedure, assembly of component parts, fixing to house, etc., all required considerable labor and expense. A local distributor was subsequently called in to assist the contractor in erecting the structure."

Based on the initial results of produce growing, some suggestions on greenhouse management are made in the paper.

Figure 139

"Media for this research was one-half peat and one-half vermiculite. Slow release fertilizers used were adequate and desirable since the gardener is relieved from concern over fertility. Rates were 5 lbs. magamp and 10 lbs. osmocote per cu. yd. Subsequent fertility needs should be added with soluble fertilizers based on visual observation, although none were used in these tests. All beds had adequate drainage to permit open system hydroponics or drainage of excessive water. The peat-vermiculite media is quite light and lacks porosity. The media will be amended in the future with finely ground pine bark. A suitable ratio seems to be 1/3 peat, 1/3 vermiculite, and 1/3 pine bark. There is concern over the expense of this media. The initial cost exceeded $200. It is proposed that the occupants of the residence prepare medias from more inexpensive materials. Leaf compost (made at home), pine bark, rotted sawdust and builder's sand are local choices. Other areas in the southern United States might have rice hulls and other choices.

"Management flexibility of any home system should allow the operator freedom to travel or leave home for extended periods. This was accomplished by combining an automated watering system with the slow release fertilizers. The system is simple and inexpensive, less than $100. It consists of a 7-day time clock, electric solenoid valve, pressure regulator, pressure gauge, lateral piping and trickle irrigation tube. Two trickle lines were placed in each bed and the system operated at 4 lbs. pressure. During cold winter months, operation time was only 30 minutes per week. Hanging baskets were also watered with the system.

The authors felt, at the time of the paper's publishing, they had insufficient data to accurately predict performance through the entire heating season. During one week in February, acknowledged as "sunnier than normal," the solar system accounted for 58% of the total heating with the house maintained at 68–70°F. They stressed in the conclusions of the paper that the pre-fab greenhouse kit was too expensive and labor intensive for its benefits. A new greenhome under construction in the program is to have the greenhouse built on the site.

Specifications

Area: 1,917 ft^2 (176 m^2)
Orientation: Due South
Wall: 1 x 8 (25mm x 200mm) yellow pine boards, rough-sawn, 2 x 6 (50mm x 150mm) studs at 16" (400mm) o.c., ½" (13mm) gypsum board finish, 6" (150mm) batt insulation
Glazing: Wood frame wall, double glazed windows
Floors: 1st floor - 2 x 8 (50mm x 200mm) floor joists at 16" (400mm) o.c. with 4" (100mm) batt insulation with vapor barrier faced down, ¾" t & g plywood flooring, various floor finishes
Roof: 2 x 6 (50mm x 150mm) wood truss, ½" (13mm) plywood sheathing, building paper, asphalt shingles, 12" (300mm) batt insulation

Greenhouse

Gross Area: 396 ft^2 (36.4m^2)
Growing Area: 134 ft^2 (12.3m^2) bench, 58 ft^2 (5.3m^2) bed plus 80 ft^2 (74m^2) porch area
Type: Proprietory lean-to greenhouse structure aluminum framed, fixed to a raised concrete block greenhouse foundation wall
Glazing: Double-strength greenhouse "B" single glazing on three sides. Continuous lapping glazed ridge vent.
Ventilation: Wet pad evaporative cooler with capacity of 5500 cfm. A 2 ft^2 (600mm^2) automatic wall shutter at each gable end, thermostatically controlled. Wood door at each end of greenhouse porch. Continuous ridge vent.

Collector

Type: Air-on-site construction
Gross Area: 576 ft^2 (53m^2)
Tilt: 60° off horizontal, oriented due south
Construction: 23¼" x 48" x ⅛" (.6m x 1.2m x 3mm) thick, low-iron, tempered glass panels (total of 72 panels—, 1" (25mm) air space, 5 mil weather-resistant composite polyester film, 3½" (88mm) air space, tempered rib aluminum roofing painted flat black, 6" (150mm) batt insulation.

Storage

Type: Washed railroad storage ballast—2" minus (50mm)
Weight: 94 lb/ft^3 (1505 kg/m^3)

Heat Capacity: 19 BTU/ft^3/^{o}F (1274 kJm^3 ^{o}C)
Volume: 1820 ft^3—85 tons (1391m^3).(86.3 tonnes)
Container: Concrete block foundation enclosures with 2'' (50mm) perimeter rigid insulation, 2'' (50mm) rigid insulation beneath northern 1/3 of rock bed, 1'' (25mm) insulation beneath middle 1/3, and no insulation beneath southern 1/3 of bed respectively, vapor barrier under entire bed.
Auxiliary: Warm air type system with a 3½ ton air source heat pump, 10kW strip heating

Hot Water

Collector type: ¾'' (19mm) copper pipe with 2'' (50mm) square continuous aluminum fins
Length: 70 linear feet (21m)
Storage Site: Two-tank system. Total of 94-gallon (365 liters) capacity—42 gallon (159 liters) tank with electric element, 52-gallon (197 liters) glass-lined tank.

Controls

Two-stage heating and cooling thermostat for solar distribution and auxiliary system differential thermostat for heat collection control. Set point thermostat—outdoor air, cooling control.

Ecotecture Group, Cedric Green, University of Sheffield, England

The British are the masters of experience in designing, building and managing greenhouses; or to use the term they created, conservatories. Sir Joseph Patton's creations, like the Crystal Palace, still evoke memories of great curving arches supporting a glass fairyland. J.C. Loudon's studies of the effects of beam radiation on panes of glass are the foundation of today's passive solar work.

The climate of the British Isles is distinctly different from most parts of the U.S. First, the percentage of diffuse overcast sky conditions is considerably higher than almost anywhere in this country. This, combined with high latitude (and correspondingly low sun angles and very short days in winter) makes a high solar fraction in December and January an impossibility. However, there are compensations which favor solar. While the cloudy, wet weather makes for extremely cold *feeling* conditions, the actual degree days are considerably less than in the northern United States. Another factor that favors solar, and

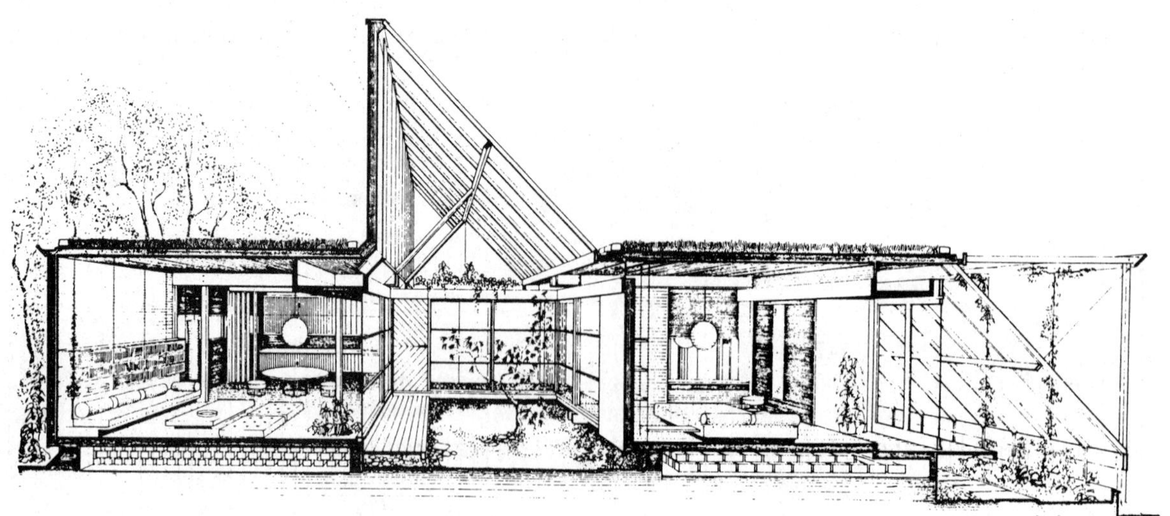

By Cedric Green

Figure 140

1. Hot air in the conservatory rises and enters the horizontal duct at the top of the heat store
2. When the temperature of this air is hotter than the bottom of the heat store, a fan pushes it down the vertical heat store. As it passes through the heat storage material the air gives up its heat
3. Another fan assists the flow of air through the horizontal heat store back into the conservatory after giving up its heat
4. At the bottom of the heat store the air is free to enter the air space between the vertical glazing and the black metal sheet. The hot air rises into the top of the heat store, sucking into the space more air from the bottom of the heat store by thermosiphonic action
5. Flaps at the top of the heat store may be opened by occupants to release hot air to the living space. This may be fan assisted at night
6. Low cost flexible solar collector under conservatory glazing connected to first hot water tank. The system may be extended to cover the length of the conservatory during summer, removing excess space heat to heat domestic water
7. Cooker fumes recirculation unit
8. Waste hot water holding tank under floor of services unit
9. Warm air from top of court conservatory is drawn down with the hot air produced in vertical high level collector and pushed into the heat store under the floor
10. Electronic 'black box' automatically switches fans and flap mechanisms according to temperature differences in spaces and stores

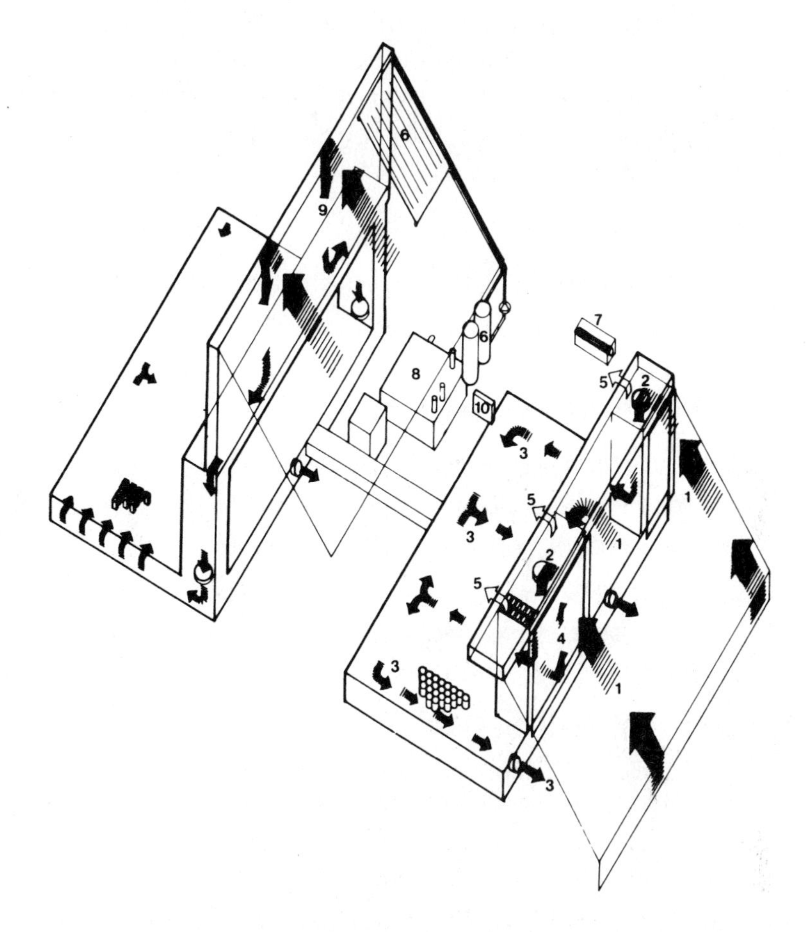

Figure 141

particularly greenhouses, is a year round need for heat in many locations. I was in Scotland in late July and nearly froze to death. Everyone was wearing wool sweaters, and the oil burners were going in all restaurants and homes.

The design shown here, takes into account these two considerations. Notice that the glazing is tilted at a 45° angle. This is an angle that favors both diffuse sky collection and heating over the course of an entire year. The building has the section on the right complete. The center conservatory and the remainder of the house on the left are to be completed shortly. Initial observations indicate that the conservatory/residence should obtain over 60% of its heat from the sun throughout the year (Fig. 140).

The entire project at Sheffield involves much more than a solar greenhouse. It is intended to demonstrate the feasibility and attractiveness of a style of living in an urban context that uses a minimum of now-renewable resources in long term consumption and construction.

1. RESOURCE AND ENERGY CONSERVATION

The approach adopted has resulted from a critical evaluation of the economics of energy conservation systems with the conclusions that the most cost-effective and adequately efficient do not need complex mechanical technology, but are semi-passive naturally and structurally integrated methods aided by simple electronic controls.

Consumption of energy is reduced to the minimum by:

a high standard of insulation;

b ventilation control by weatherstripping and using plant-filled conservatories as "filters" and air-locks;

c solar energy collection using air as a collection and transfer medium, small containers of water and fireclay blocks as a short term storage medium, conservatories as a low cost means of providing a large area of glazing, and plants as solar converters, filters, humidifiers and food;

d solar water heating in summer diverting heat gain by means of low cost "blind" collectors shading the interiors.

Consumption of water is minimised by:

e collection and storage of organically filtered rain water for washing and watering;

f reuse of washing waste water for wc flushing;

g use of shower and spray taps;

h solar distillation to produce drinking water.

Constructional conservation is achieved by:

j use of timber as a basic construction and finishing material;

k use of recycled or processed waste materials;

l collection of reject materials and containers (for water);

m elimination of processes involving heavy mechanical plant;

n designing for construction by hand processes aided by small power tools.

2. ECONOMY OF CONSTRUCTION

Related closely to resource conservation, capital costs have been reduced to the minimum to allow extra capital costs of the energy conservation system to be paid back by energy savings in the shortest posible time viz. 7 years at present rates of fuel cost inflation in the area of study. This represents 19% p.a. return on capital investment after deduction of mortgage interest.

Savings in capital cost are achieved by:

a integration in construction of all ducts, heat storage enclosures, etc.;

b provisionof solar collectors as useful space in the form of conservatories with consequent savings in circulation and heated areas;

c standardisation and repetition of components and rationalised dimensions to use standard off-the-peg materials and components;

d off-site component prefabrication of complete frame and panel construction, rapid site erection and creation of weatherproof enclosure;

e potential for owner-builders, either buying complete kit-of-parts or doing their own prefabrication from detailed manual, thus making further construction cost economies by providing own labour.

f simplification of planning details and careful calculation to minimise spans and member sizes;

g use of insulation as structural core of sandwich panels;

h small capacity back-up heating system resulting from lowered thermal consumption.

Sundwellings Demonstration Center, Ghost Ranch, Abiquiu, New Mexico

The Sundwellings Project at Ghost Ranch is an opportunity to test three passive solar applications (direct gain, Trombe wall, solar greenhouse) against a control unit. The project brought some of the best known names in New Mexico solar design together under Peter van Dresser's coordination. Although modest in funds, the Sundwellings program has had major impact as a model for solar training. It included the following goals:

1. A Training Program to train residents of Northern New Mexico in the techniques of design and construction of solar heated dwellings and solar water heaters. A 16-week program involving sixteen trainees consructed two of the test units. CETA employees from Santa Clara and Dulce along with Ghost Ranch employees from Abiquiu completed the construction phase.

2. A Demonstration Center open for permanent public inspection and education, the solar units demonstrate how low-cost passive sundwellings can be constructed. Use of the rooms by guests at Ghost Ranch will give ''live-in'' experience of the energy-conserving performance of the systems.

3. A Solar Testing Facility The buildings are identical in size, construction, and compass orientation. Only the systems of heat collection and heat storage differ. Instruments and monitoring equipment installed by trainees under the direction of the Los Alamos Scientific Laboratories will allow precise evaluation and comparison of the systems. This is the first facility like this in the world.

I was involved in the design of the greenhouse unit and ran the workshops that put it up. At the time, I optimistically predicted that the solar greenhouse would outperform the other units. (I had a vested interest, of course.) The technical consensus was against me. It was predicted that the direct gain would be first, Trombe wall second, and the greenhouse third. After two winters of testing and analyzing, I'm happy to say my prediction came true. The greenhouse does approximately 80% of the heating for its units. (The direct gain is second, and the Trombe wall is third.) I'm not too smug, however. The greenhouse has about 30% more collector area than the other units.

This larger area for solar collection is a primary advantage a greenhouse has over other passive solar applications.

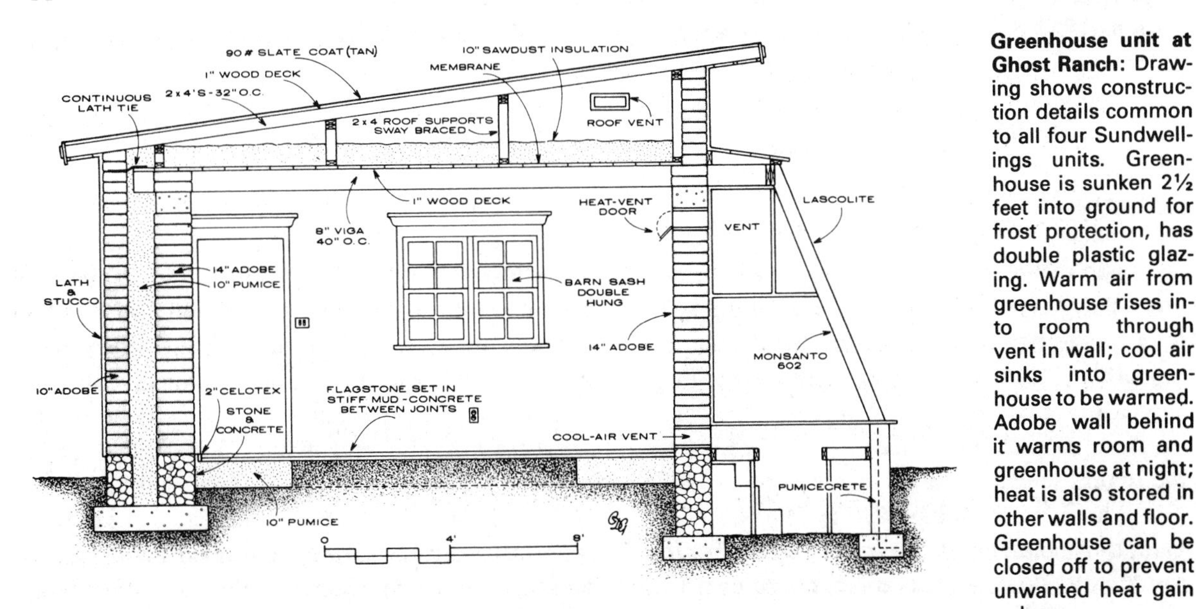

Greenhouse unit at Ghost Ranch: Drawing shows construction details common to all four Sundwellings units. Greenhouse is sunken 2½ feet into ground for frost protection, has double plastic glazing. Warm air from greenhouse rises into room through vent in wall; cool air sinks into greenhouse to be warmed. Adobe wall behind it warms room and greenhouse at night; heat is also stored in other walls and floor. Greenhouse can be closed off to prevent unwanted heat gain or loss.

Figure 142

Reprinted with permission from Solar Energy Handbook, *published by Popular Science Time Mirror Magazines, Inc.*

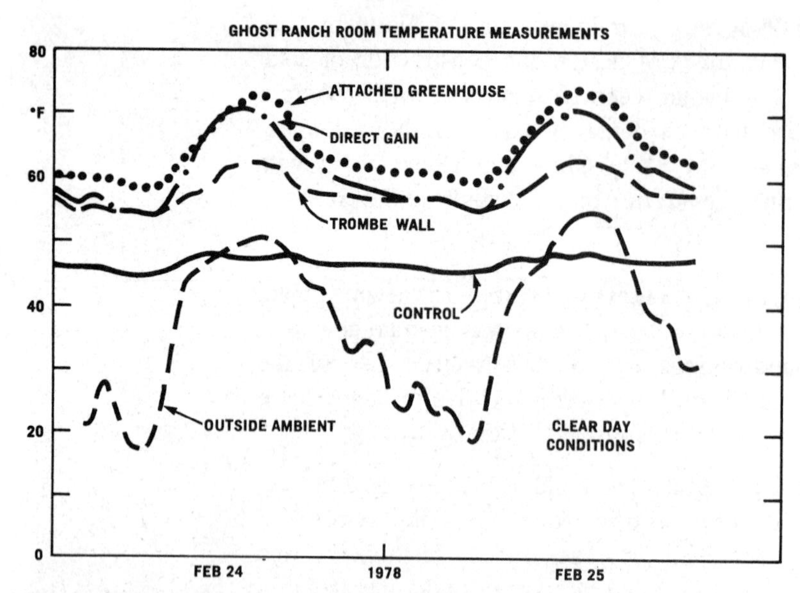

How temperatures compare in the four Sundwellings dormitories at Ghost Ranch, in the mountains of northern New Mexico, during two clear February days: The buildings were all unoccupied and totally unheated—except by the sun—during this period. Temperatures were measured inside the four buildings—greenhouse, direct gain, Trombe wall, and control—with globe thermometers.

Figure 143

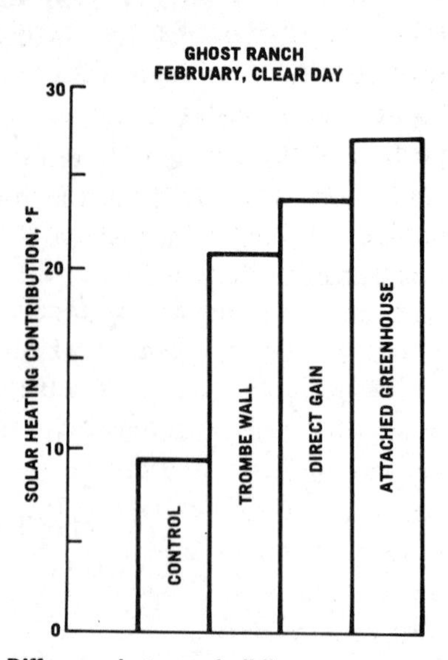

Difference between building temperature and average outside temperature is one measure of solar-heating performance. The bar chart above shows this difference for the four Ghost Ranch buildings on a clear winter day. Here the solar greenhouse does best, despite having larger losses through the glazing, because it also receives much more sun through the larger glazed area.

Figure 144

Ekose'a, 573 Mission St., San Francisco CA 95105

One energy conserving structure receiving wide attention is the Tom Smith house which overlooks Lake Tahoe from the California side. The house was designed by Lee Porter Butler and built by a local contractor at $28.50 per square foot, less than local construction cost levels using conventional building designs ($37.00/sq. ft.). The house is approximately 1,800 square feet (includes 330 square feet of greenhouse space) and has three bedrooms, two baths, a kitchen, and combined living-dining area.

The house is 100 percent passive, utilizing no fossil fuels, electricity or mechanical systems for comfort maintenance, and maintains comfort levels of temperature and humidity year-round. The contractor employed conventional building methods with readily available building materials to construct the home in the style preferred by the owner. The house meets or exceeds all codes, health/safety standards and financing requirements.

The technology itself is elegant and simple. The south side of the house is a greenhouse/solarium area. The greenhouse is part of the active air space extending not only over the glazed openings, but also up over the ceiling in what would otherwise be dead air space, through the attic, down the opposite wall (usually the north wall) and underneath the floor through the crawl space back around to the greenhouse. This forms the complete 'thermal envelope' within which air is free to circulate back and forth depending on the relative density of the air (due to temperature differences) in the vertical areas of the envelope. Thus, the main living areas are surrounded by a gently moving circle of air.

The flywheel or inertial effect in the thermal envelope works by air convection. During times of solar gain, the cooler, denser air on the north side of the house falls forcing the warmer, expanded and lighter air in the greenhouse to rise and circulate over the ceiling space to be cooled again on the north side. The circulating air affects and is affected by the surface temperatures of the building materials and earth mass which make up the interface of the thermal envelope.

At night, or during periods of heat loss, the flywheel effect works in reverse as the south face glazed greenhouse cools faster than any other portion of the thermal envelope. The earth mass then serves to warm the cooler air falling out of the greenhouse, this warmer air circulating up the north face and across the ceiling, once again to fall through the greenhouse.

Figure 145

During periods of solar gain, the mass of the house and earth storage system absorb excess heat. During periods of heat loss, the mass of the system releases stored heat to the active air space which minimizes temperature effects on the interior living areas.

Tom Smith reports that the experience of living in this house is very different from conventional houses in that the air is fresh, rich and warm, there are no drafts and no cold spots in the house. Some other benefits include the constant and healthful humidity level of 50% to 60% year-round, the absence of dust, the quietness of the house, and the livability of the greenhouse area (which can be used about 75% to 80% of the time year-round).

While it may be easy to understand how and why the house works in the winter, it is equally important to realize that the house successfully cools in the summer, also without the use of backup systems. This is accomplished by the dual effect of venting warm air out of the upper portion of the greenhouse space and drawing cool air into the crawl space by means of an underground duct on the north side of the house. (See Fig. 149, p. 134).

Figure 146

A case history analysis of the house is available from Tom Smith and Lee Porter Butler for $18.95, prepaid. Ekose'a also offers a Preliminary Planning Package which includes ten preliminary plans for houses located in four widely differing climatological regions of the United States and Canada, explanation and illustration of design concepts, principles and methodologies, a checklist of design considerations, presentation of the type of modifications necessary to build a standard plan Ekose'a house in another climatological area, glossary of terms, and answers to frequently asked questions. The price of the Preliminary Planning Package is $65, prepaid. Send payments for any desired materials to: Ekose'a, 573 Mission St., San Francisco, CA 94105.

The diagrams shown refer to a generic design rather than this particular house.

During the day (Fig. 147) the sun enters the greenhouse/solarium south envelope space heating the air. At the same time, cold outside temperatures cool the air in the north envelope space. The colder air is heavier than the warm air. The heavier air is pulled down by gravity pushing the earth-tempered air in the crawl space up and into the greenhouse space, which pushes the warmer air in the greenhouse up and over the attic space.

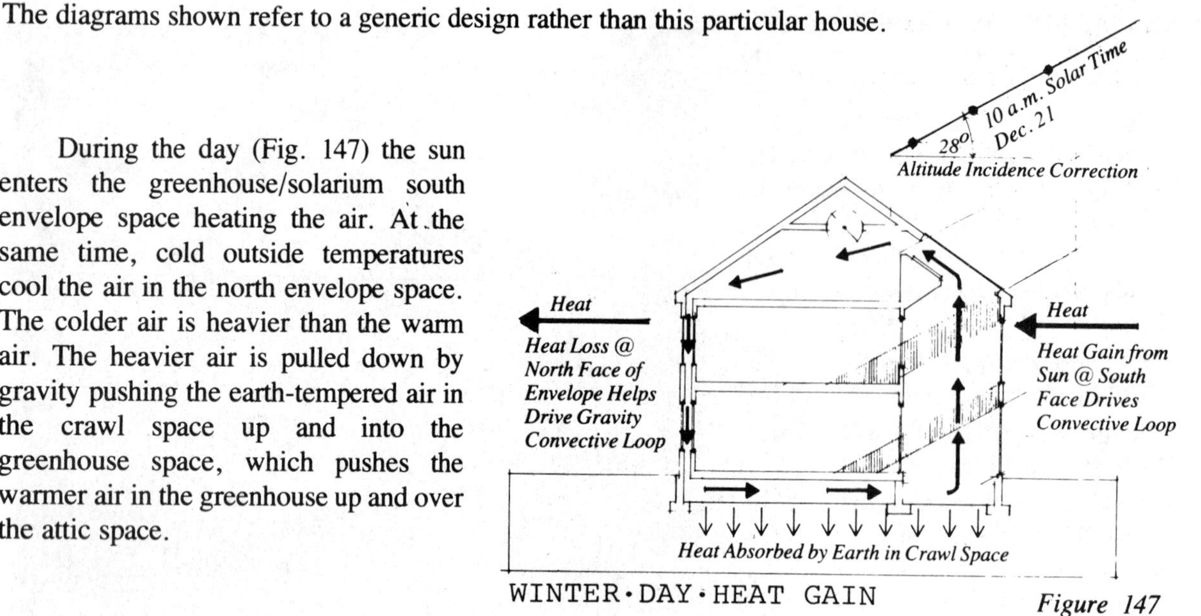

WINTER•DAY•HEAT GAIN

Figure 147

At night (Fig. 148), the greenhouse loses heat faster than other parts of the envelope because of the large glass areas. This colder air falls into the crawl space where it is warmed by the earth and then pushed up the north envelope space and over the ceiling and back into the top of the greenhouse.

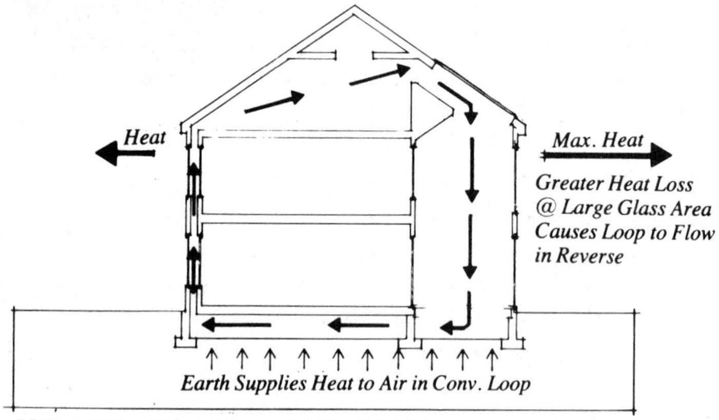

Heat

Max. Heat

Greater Heat Loss @ Large Glass Area Causes Loop to Flow in Reverse

Earth Supplies Heat to Air in Conv. Loop

WINTER•NIGHT•HEAT LOSS

Figure 148

During hot summer conditions (Fig. 149), the roof prevents the sun from striking the ceiling, thus reducing heat gain. East and west glazing is held to a minimum. The attic space heats up as usual, but since the air is allowed to escape, there is a constant flow of cooler air entering the house through the use of underground pipes.

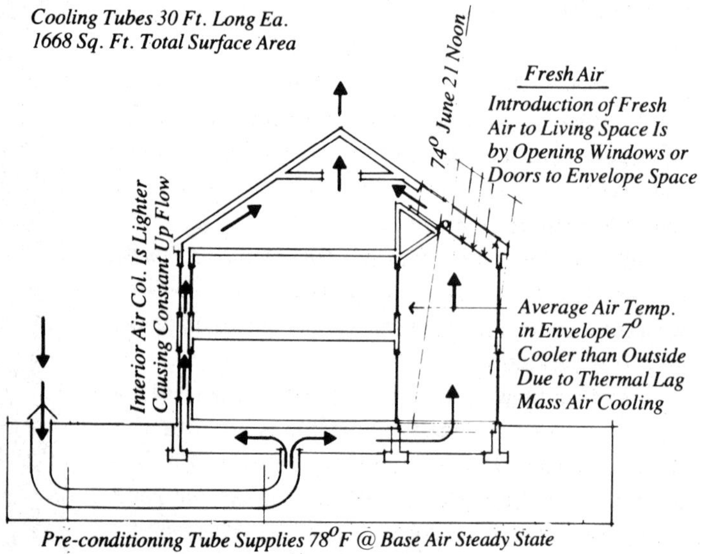

ENVELOPE VENTILATION & COOLING

Figure 149

Rick's Cafe, Solar Power Supply, Inc., Denver, Colorado

The greenhome concept is being used in commercial applications. This is Rick's Cafe in Denver (Fig. 150) where President Carter had dinner on National Sun Day in 1978. The design work was done by Solar Power Supply of Evergreen, Colorado. An innovation used here is called the neutral density collector. The interior panels of the upper double glazing is a gray-colored plexiglass. The darker color of the interior plexiglass absorbs radiation and causes high air temperatures in *between* the glazing. The heat is then taken out of the glazings at the apex and blown down to rock bed storage under the floor of the cafe (Fig. 151). While this is not as efficient as using regular double glass in handling the warm air inside the greenhouse, it was done for a specific purpose here. The cafe has patrons all through the late morning and early afternoon period throughout the entire year. By reducing solar transmission, the inside light is softer and temperatures are less extreme. It's a good idea for a greenhouse which has continuous daytime occupancy.

Figure 150

Helion, Inc.
Box 445, Brownsville
California 95919

1. Solar Greenhome 1 is the first of several greenhomes to be constructed by Helion. A series of design studies paved the way to a simple structure—primarily post and beam/blockwall. Design work was done by Jack Park, with input from Marianne Langhorn and Helen Park.

2. A solar greenhome—the term started at Helion—is the best combination of dwelling structure and solar greenhouse for climate control and food production.

3. Principal solar heat gain is from greenhouse glazing to the 16'' thick Trombe heat storage wall which doubles as a structural wall. Secondary heat gain is by optional glazing on the sloping south wall of the loft. Helion has chosen insulated sliding glass windows for the first version of this dwelling to enhance summer ventilation—mild summer climates will not demand additional ventilation.

Figure 151

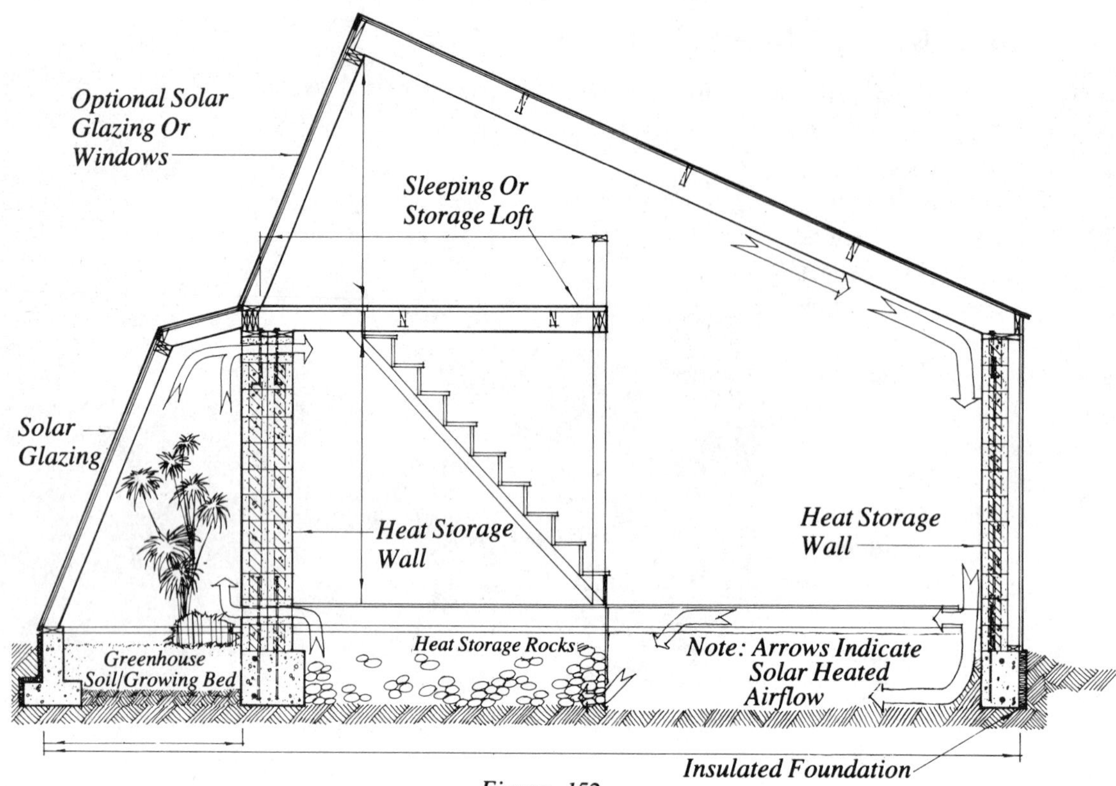

Figure 152

4. Secondary heat storage is produced by the structural north block wall—with the insulated curtain wall outside of it—and (10) ten or more tons of fist-sized washed river rocks under the floor. Primary air flow is through these rocks into the greenhouse.

5. An optional wood deck allows construction of Greenhome 1 on a sloping site, as was Helion's prototype.

6. A loft over vestibule provides a small guest sleeping quarters or additional storage space.

7. Roofing material is corrugated aluminum sheets cut to just under 21'0'' long for the main roof, 2'-0'' long for the greenhouse and 12'-0'' long for the vestibule. The steep slopes of these roofs makes this material difficult to install, but with careful installation this yields a lifetime, fire resistant roof.

8. All vents penetrate sidewalls, rather than the roof membrane.

9. Greenhome 1 is a flexible, expandable structure, designed for an absolutely minimum cost, thermally efficient dwelling.

Helion sells blueprints for Greenhome 1. Write to the address above.

Hamilton Migel, Santa Fe, N. M.

Hal Migel is a designer and builder of exceptional talent. I first became aware of his professional skills seven or eight years ago when every adobe mudslinger I knew was saying, ''Have you *seen* the detail work in Migel's newest house?'' Attention to detail, function and form personifies his work.

Migel was the first person in this region to demonstrate that a greenhouse could be the primary heating unit for a large home (2500 square feet). He did it with some revolutionary innovations.

The double-walled clear face is separated (or inflated) by a small fan and is supported by thin steel ribbing. The entire clear wall can be detached from the front of the house in late spring, leaving a lovely outdoor garden in full bloom. Removing the cover also provides a clear view to the outside (also see Solar Room, p. 157).

Warm air is ducted from the greenhouse apex to a rock storage bed under the house, then back into the unit to complete the loop. Radiant heat from the concrete slab over the rock bed warms the house. The system provides approximately 80 percent of home heating.

Figure 153

Owner: Hamilton and Candice Migel
Designer: Hamilton Migel
Builders: Bingham and Migel, Inc.
Floor Area: 400 square feet (greenhouse)
Clear Area: 500 square feet (greenhouse)
Home Floor Area: 2500 square feet
Thermal Storage: Rock beds under floor of home
Air Circulation: 1100 to 2200 cubic feet per minute (see intake vents, upper left, Fig. 154).
Main Function: Home heat, vegetables and flowers

Figure 154

Independent Greenhouses

In this section we'll look at freestanding solar greenhouses, both large and small, and at some of the work going on in research institutions to make conventional greenhouses more energy efficient and productive. It's important to realize that the criteria for success in these greenhouses are quite different. The home owner will desire a small structure, usually between 100-200 sq. ft., that will allow for gardening throughout the year with a minimum of expense and upkeep. The commercial grower, on the other hand, is in it as a profit making business. To date, commercial growers have been unwilling to spend vast amounts of money making their greenhouses more energy efficient. Greenhouse farming has never been a big money maker, and current uncertainties in fuel availability make their reticence easy to understand. So, any improvement in energy conservation they might make must be weighed against capital investment, depreciation and a long list of other factors that are found in any business. The newest entry in the solar greenhouse field is the community greenhouse. This is an exciting concept to which we will devote part of this section.

Brace Institute, McDonald College of McGill University, Quebec, Canada

Tom Lawand and his colleagues at Brace Institue were dealing with a specific problem when they began their pioneering efforts in solar greenhouses in early '70's. In their locale are many small farmers trying to earn a living by truck farming vegetables and with seasonal sales of flowers and seedlings. The area is characterized by both short growing seasons and killing early frosts. The institute knew that small greenhouses could greatly aid these farmers but didn't want to substitute one problem for another by recommending energy-intensive conventional designs. The outgrowth of these challenges was the design shown here. It is apparent by the relatively small amount of glazing that this is a design specific to a high latitude and limited growing season. However, the principles of good thermal greenhouse design began by Brace have evolved and spread throughout the solar greenhouse field. Besides optimizing the tilt of the north wall and their work with reflective interior walls, Brace has recently been involved with materials and mounting systems for movable insulation in the greenhouse.

Much of the Institute's effort is directed toward appropriate technologies in third world agriculture and housing. More information on their publications and blueprints of these greenhouse are available by writing them. Inquire as to the cost of the publications and include a S.A.S.E.

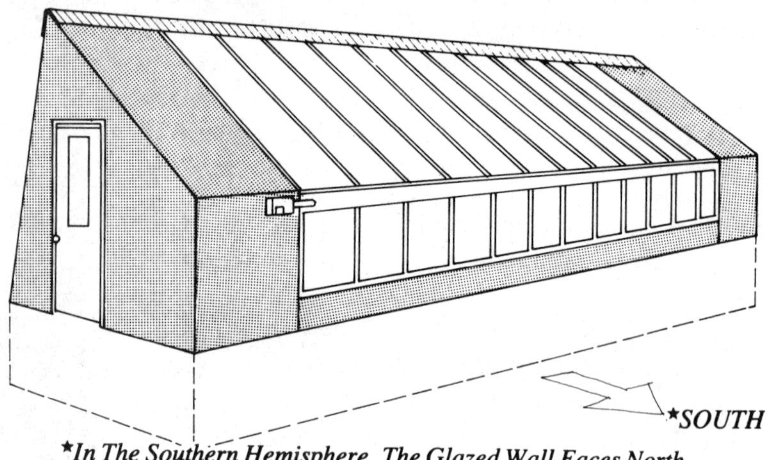

**In The Southern Hemisphere, The Glazed Wall Faces North*

Figure 155

The Brace Greenhouse—General Principles

- The peakline is oriented east-west.
- The north wall is insulated to reduce heat losses.
- The north wall is inclined toward the sun at an angle of 68°, chosen to avoid shadows at the summer solstice *at the 47° latitude*.
- The south roof is glazed and inclined at 35°. This angle is not optimal for capturing solar radiation but forms a configuration that is convenient for the operator who must stand and work within the structure, yet avoids the large volume of air that would have to be heated in a higher peak.
- The interior of the north wall is covered with a reflective surface, which serves to increase the amount of radiation reaching the plant canopy.

Reed E. Maes, Environmental Research Institute, Ann Arbor, Michigan 48107

The innovations found in the Maes greenhouse could have long range repercussions for cold climate food self sufficiency. Reed Maes who conceived, financed and constructed this 6000 sq. ft. structure on his own, had stringent criteria for the operation. First of all, he feels that it is mandatory that northern states reduce their dependency on outside sources for their fresh food in winter. Next, in order for the concept to be viable for the small farmer and grower it had to be low in cost initially and easy to manage. Last, Reed knew that a small, commercial operation could not afford to pay the heating bills a conventional greenhouse in this severe climate area would incur and stay cost-competitive with Southern California and Mexican produce.

The result of his labors in this structure are innovative in both technical features and management.

Figure 156

"The winter garden is 58 meters long in the east-west direction to take advantage of the winter sun and is 9.1 meters wide. It has an area of 532 M^2 or approximately 6000 ft^2. An A-frame is basically 2 x 4 wood construction using 24 ft. lengths for the long triangular members. The lower 8 feet contains most of the framing, while the upper level only supports the polyethylene cover and the venting mechanism. The lower third of the south wall used transparent fiberglass on the inner surface, and when this is coupled with the outer layer of polyethylene it results in a loss factor of about R-1.5. The ceiling is constructed of 2 inch thick foil-faced polyurethane sheets that have an R value of about 16. The north wall uses standard construction materials and also has a loss value equivalent to R-16. Therefore, the only high thermal loss surface of the garden is the 8 ft. high south wall. This configuration results in heat losses less than 25% of a similarly sized conventional greenhouse.

"In the low energy greenhouse it is also necessary to reduce ground losses around the perimeter of the structure to a low level. This insulated footing is especially important because the ground is the important thermal capacity of the garden and the main source of stored heat. The 5 cm styrofoam insulated footing was installed below the frost line to a depth of 1.2 meters. Water drum and gravel trench storages were also employed to add to the thermal capacity of the garden. The 200 liter drums were placed along the north wall, whereas air was circulated through the rock storage with a number of circulating fans placed at intervals extending the length of the garden. The gravel storage was found to be very inefficient and therefore the fans have been used very little.

Figure 157

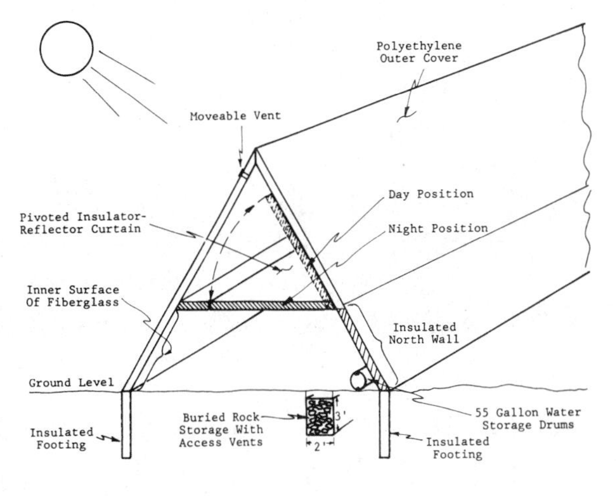

Figure 158

Figure 159

"The severe Michigan winter of 1977-78 was a good test of the structural advantages of an A-frame. Snow loads during the winter were shed from the steep slopes with no melting heat required from the interior. Also, the wind loads and blizzards were survived without mishap.

"The garden areas were planted with lettuce, broccoli, radishes, garlic, onions, cauliflower and spinach in October. These crops matured during the winter with no added heat, and second crops of lettuce and spinach were harvested. Venting is required on most sunny days because the excess heat cannot be adequately stored.

"For test purposes, the greenhouse was divided into two 29-meter sections in mid-February. It now has a cold weather section and a section heated by a furnace to maintain a temperature of 62°, adequate for warm weather crops such as tomatoes and cucumbers. The tomatoes and cucumbers ripened in early May. The heat losses have been monitored since the heat was turned on. The degree-days during this time frame were 3500 and the heat required was 76 KWHR/M^2 or 12,000 BTU/ft^2. This projects to about twice this amount on a yearly basis and is very modest compared to conventional greenhouses."

In addition to his technical innovations Reed has developed a marketing strategy for the produce. Because all of the vegetables are organically grown, the production is contracted through a local food co-op. So, local residents can get organic home-grown food throughout the year in Ann Arbor.

Figure 160

The Herb Shop

This unit might be considered the mother of solar greenhouse design (Fig. 160). As far as I know, it was the first such commercial greenhouse in the United States. The design is based on the work of the Brace Institute (in Çanada) and T.A. Lawand. Joan Loitz, the owner, adjusted the angle of the front face and the reflective back wall to correspond to the latitude of Santa Fe, New Mexico. The north wall is at a 78° angle (angle of the sun at summer solstice) and is insulated with 4'' of fiberglass.The interior is paneled with Masonite and painted with a glossy white enamel. The rafters extend from the roof peak to the ground and were originally uncovered for the lower 10 feet. Joan covered this area with corrugated Lascolite to make a greenhouse "preheater" or "buffer zone" and cold frame area. It stays about 25 degrees colder than the main greenhouse at night. This addition cost her 90 cents per square foot and gave her an extra 480 square feet of growing space.

Occasionally a customer will challenge Joan about her use of supplemental gas heating. She replies that she's in the business, not playing games, and that the heater is the most cost-efficient way of backing up a system that's 80 percent solar. Joan estimates that it would cost her $3,000 to install an active solar system to capture the 20 percent non-solar power that the gas company furnishes. Her average $32 a month gas bill for the eight-month cold season compares to $160 a month for a similarly sized conventional greenhouse in the area, and bears out the effectiveness of her design. Incidentally, she's running 70 degree nighttime lows, which is considerably higher than those maintained in non-commercial greenhouses.

Some thermal storage is provided by sixteen 55-gallon drums that also double as table supports. Although these drums receive little direct sunlight, they do contribute to heating when the ambient air drops below their temperature and help to stabilize temperature fluctuations.

One drawback to the design of this unit is that the entire clear area approximates a normal angle to the summer sun. This may be all right for Quebec, but it creates overheating and plant-burning problems in Santa Fe. Joan has used some ingenious low-cost methods to combat this problem. One of them is to attach a $1.98 misting nozzle to a hose and hang it in front of the air circulating fan. This will lower the temperature 10 degrees in 10 minutes; she turns it on several times during a summer afternoon. She also trains scarlet runner beans up the west wall to block some of the afternoon sun.

Joan and I have often worked together in planning community solar greenhouse systems. Her particular expertise is in organic growing methods and biological control of insect pests (see Chapter VII). She ships 100 varieties of organic solar-powered herbs all over the world. You can order from the address given below.

Owners: Joan Loitz, The Herb Shop, 1942½ Cerrillos Rd., Santa Fe, NM 87501
Designers: Brace Institute of McGill University, and Ms. Loitz
Builders: Joan Loitz and friends
Floor Area: 1600 square feet
Clear Area: 1042 square feet (greenhouse); 480 square feet (cold frame) = 1522 total square feet

University of Arizona Environmental Research Laboratory

This research center is concentrating on the development of integrated systems that provide power, water and food. Staff members at the laboratory are experimenting with aquaculture and vegetable production within polyethylene-covered greenhouses. Plastic-lined tanks contain shrimp that feed on organisms living among water hyacinths in the pond (Fig. 161).

Figure 161

An interesting secondary project that is being developed is an insulating greenhouse cover similar to the Beadwall[R] that injects liquid foam between the plastic layers at night (see Fig. 45).

Another facility created by the Environmental Research Laboratory expands the ecosystem approach to include desalination of sea water and nutrient preparation for the greenhouse use. This ambitious project is being carried out in collaboration with the University of Sonora and is located at Puerto Penasco, Mexico. In this design, waste heat from engine-driven electric generators is used to desalt the sea water.

The fresh water is then piped to vegetables within controlled-environment greenhouses of air-inflated plastic. The researchers state:

> "The concept is applicable to vast regions where almost nothing grows—and where desalted water remains prohibitively expensive for open-field agriculture. The principal advantage of the concept for arid regions is extreme conservation of water. Moisture lost from field crops, by evaporation and transpiration, is enormous, of course. In a closed system this moisture can be captured. Estimates are that a plant within such a sealed-in environment uses only about a tenth as much water as it would need outdoors."

Energy Conservation and Solar Applications to Conventional Large Scale Greenhouses

Commercial growers who own conventional greenhouses are in trouble. Besides having to compete with produce shipped across the country (or from other nations) they must try to stay competitive while paying huge fuel bills on energy gobbling greenhouses. Many of the larger growers, who measure the area under the glass in acres, are family operations owned and managed by the children or grandchildren of the people who built these durable units originally. It's ironic that structures of this quality of construction also represent such ineptitude of thermal design.

Luckily, there are people who have chosen to tackle the most difficult of greenhouse problems: making it technically and economically feasible to heat these dinosaurs.

Ohio Agriculture Research and Development Center, Wooster, Ohio

Here are some excerpts from "Conserving heat in glass greenhouses with surface-mounted air-inflated plastic" by W.L. Bauerle and T.H. Short.

"Steam flow meters and recorders were placed in both the plastic-covered glasshouse compartment and an identical conventional glasshouse compartment to measure the total amount of energy used in each. Solar radiometers were placed inside and outside of each compartment to measure sunlight accumulation. Wind measurements were recorded to determine the total miles of wind per day across the experimental compartments. Total yield and fruit quality data were recorded.

"Measurements of energy requirements for each of the two compartments from December 1975 through March 1976 (the 4 coldest months of the year in Ohio) revealed that the double plastic cover reduced the total heat use by 57%.

"Glass and structural members of the conventional glass greenhouse reduced the solar radiation by an average of 35%. An additional reduction of 18% was measured in the double plastic-covered glasshouse. The plastic cover, however, tended to diffuse the light more uniformly within the greenhouse and structural shading was less apparent.

"The daytime relative humidity in the plastic-covered glasshouse was approximately 12% greater than in the conventional glasshouse during the winter months when the vents were closed. High levels of humidity tend to reduce plant stress and improve pollen viability on a spring tomato crop. No difference in disease incidence was observed.

"The total number of fruit on each plant was similar for both compartments. However, fruit weight per plant was reduced in the plastic covered house by 6.5% for W-R 25, 4.3% for M-R 13, and 10.3% for Hybrid O. (Hybrid O is not normally grown as a spring crop.)

"The economic implications of these results have caused much grower interest in the double-cover system for glasshouses. A few growers have already tried the system and have encountered technical difficulties with installation. Other growers have successfully used this system for 2 years.

Observations

"The plastic should remain pressurized at all times. Normal operating pressure will keep the plastic from having wind ripples and make it feel soft to a gentle push. Many growers installing double plastic for the first time become alarmed at billowing after inflation. Considerable billowing is normal and effectively allows the outer cover to adjust to variable wind and internal pressure loadings. If allowed to remain deflated for a number of days, the wind-whipped plastic will slowly begin to tear near the fasteners.

"The effect of double covers on paint peeling has not been fully assessed. However, little or no peeling has been observed with air-inflated double covers. The inside cover is pressed so securely against the bars that it may actually protect the paint. Most grower experiences with plastic covers causing paint peeling have been associated with single, uninflated covers on sidewalls. Uninflated plastic covers typically cause paint peeling by collecting moisture and hammering the paint on windy days.

"Heavy snow loading on gutter-connected houses may be a major problem. Generally, an air-inflated plastic cover tends to distribute snow loads much better than glass. Normal snow and ice accumulations melt off slowly as long as the cover remains inflated and effectively insulated. If the cover is temporarily deflated, the melting rate is greatly increased. Large snow loads tend to naturally deflate the cover in the heaviest loaded areas. The probability of glass breakage under these circumstances is not known. Neither is it known whether glass breakage would occur under the same circumstances without the plastic. It is most likely that the plastic will still maintain the integrity of the roof even if some glass does break.

"Obviously there are risks which must be taken in applying double plastic to glass greenhouses. The potential heat savings of 57% indicated by OARDC research makes the risks economically worth consideration and in some cases necessary. It is most important that growers first cover only a small area to get experience. Growers should also be sure to contact their insurance company before proceeding. The Ohio Agricultural Research and Development Center can only verify the energy and crop responses under the glasshouse and double-plastic covered glasshouse conditions.

"There have not been any engineering analyses of the structural responses to the double-plastic cover system, and OARDC cannot be held responsible for any failures.

Figure 162

"To date, no serious failures have occurred as a result of double plastic over glass greenhouses located in heavy snowfall areas. It is anticipated that further growth trials and observations will be made known to the industry for future use."

Cook College, Rutgers University, New Brunswick, New Jersey

The staff at Cook College has been addressing the problems of heating commercial greenhouses of the newer double-inflated polyethylene variety. These low cost units now comprise a major fraction of the market. The growers who use these greenhouses are often raising flowers and ornamentals that demand a much narrower temperature range than found in a home greenhouse. (65° lows and 82° highs, for instance.) For that reason, the applications are more intricate than most in this book. Cook College had the opportunity to put different systems together that had been tested seperately before in the "Kube Pak" demonstration. This 58,000 sq. ft. facility is at the Kube Pak Garden Plants in Allentown, New Jersey. Here is the system description from a paper by Mears, Roberts, Kendall and Simpkins of the Rutgers team.

"An insulation system being utilized at Kube Pak is a movable horizontal curtain which encloses the growing area at night and opens during the day to allow normal lighting conditions in the greenhouse. Substantial heat savings are obtained with properly installed single-layer thin-film curtains. Curtain insulation systems are currently being utilized with conventional heating systems to reduce heat loss and several companies are marketing systems. The system is installed at Kube Pak is based upon mechanisms previously utilized for pulling black-cloth shading systems. Curtains are pulled across each section of greenhouse by motor driven cables running just under the greenhouse gutters. Between the drive cables, located 10 ft. (3m) apart, the curtain is supported by stationary polypropylene monofilaments. As the curtain is closed, the leading edge intercepts the hanging tail of the next curtain sealing each section.

"At Kube Pak there are four new curtain materials under test: 4 mil clear vinyl, 4 mil clear vinyl laminated to 0.5 mil aluminized mylar, 6 mil black vinyl, and 6 mil black vinyl laminated to 0.5 mil aluminized mylar. During the evaluation phase of the project, these materials will be monitored and their relative heat savings and mechanical performance noted.

"The porous concrete floor system is both storage and primary heat exchanger for the greenhouse. The floor is constructed of four layers: polystyrene board insulation, a vinyl swimming pool liner, gravel flooded with water, and a cap of porous concrete. The porous concrete forms a firm floor, a heat exchange surface, and allows excess irrigation water to drain. The water provides thermal storage when mixed with gravel and is the heat transfer fluid for the solar collector and secondary heat transfer system. The vinyl liner contains the concrete, gravel and water. The insulation also serves as mechanical protection for the liner during construction.

"Temperatures in the floor are kept low to enable plants to be grown directly on the warm floor and to increase the efficiency of the solar collectors. Low storage temperatures require that a large area of heat exchange be provided. This is the function of the vertical curtain heat exchangers which are made by connecting a water distribution pipe to a horizontal support. A vinyl film is draped completely over this support. When heat is needed, a thermostat actuates a pump which circulates warm water from the floor through the water distribution pipe. As the water trickles down between the hanging curtains, the entire inside surface is wet. The water exits the bottom of the curtain and returns directly to storage through the porous concrete. Strings looped under the curtain are attached to a movable cable. When the cable is pulled, the strings roll the curtain up like a porch awning for out of the way storage overhead. These curtains are positioned over the walkways so that no plant growing space is required for their use.

"The solar collector consists of a frame covered with five layers (two clear tubes and a black sheet) of plastic film. The black plastic sheet, which is the absorber plate of the collector, is sandwiched between the two air-inflated clear greenhouse grade polyethylene tubes. The air spaces between the rear two layers and the front two layers are kept inflated with small blowers. The absorber plate is pressed between the two inflated polyethylene cushions which stabilize the structure and provide some insulation. There are three

1. AIR INSULATED GREENHOUSE
2. MOVABLE INSULATION
3. GRAVEL / WATER STORAGE
4. PLASTIC LINER
5. POROUS CONCRETE FLOOR
6. VERTICAL CURTAIN HEAT EXCHANGER
7. HEAT EXCHANGER DISTRIBUTION PIPE
8. PUMP TO SUPPLY COLLECTOR
9. PLASTIC SOLAR COLLECTOR
10. GRAVITY RETURN

SCHEMATIC CROSS SECTION OF SOLAR HEATED GREENHOUSE

Figure 163

frames under test at Kube Pak: a wooden frame based upon research prototype designs, and steel frame units utilizing greenhouse structural components. Three collector frames have steel greenhouse roof rafters previously utilized in greenhouse structures at Kube Pak. The fifth collector utilizes steel tubing used as roof supports in commercially available greenhouses. Water is introduced along the top of the collectors through perforated header pipes and flows down over the black polyethylene absorbing layer. The gutter at the bottom of the frame collects the heated water which returns to the greenhouse floor via a gravity flume. The collector orientation is south, and the slope is adjustable between 25° and 65° to compensate for seasonal changes in the sun angle above the horizon."

The technical papers of Mears, Roberts, Kendall, and Simpkins (in various order of first authorship) are often found in American Society of Agriculture Engineers (ASAE) publications and in the proceedings of the DOE-USDA sponsored Solar Heated Greenhouse Conferences.

The New Alchemy Institute, Woods Hole, Massachusetts

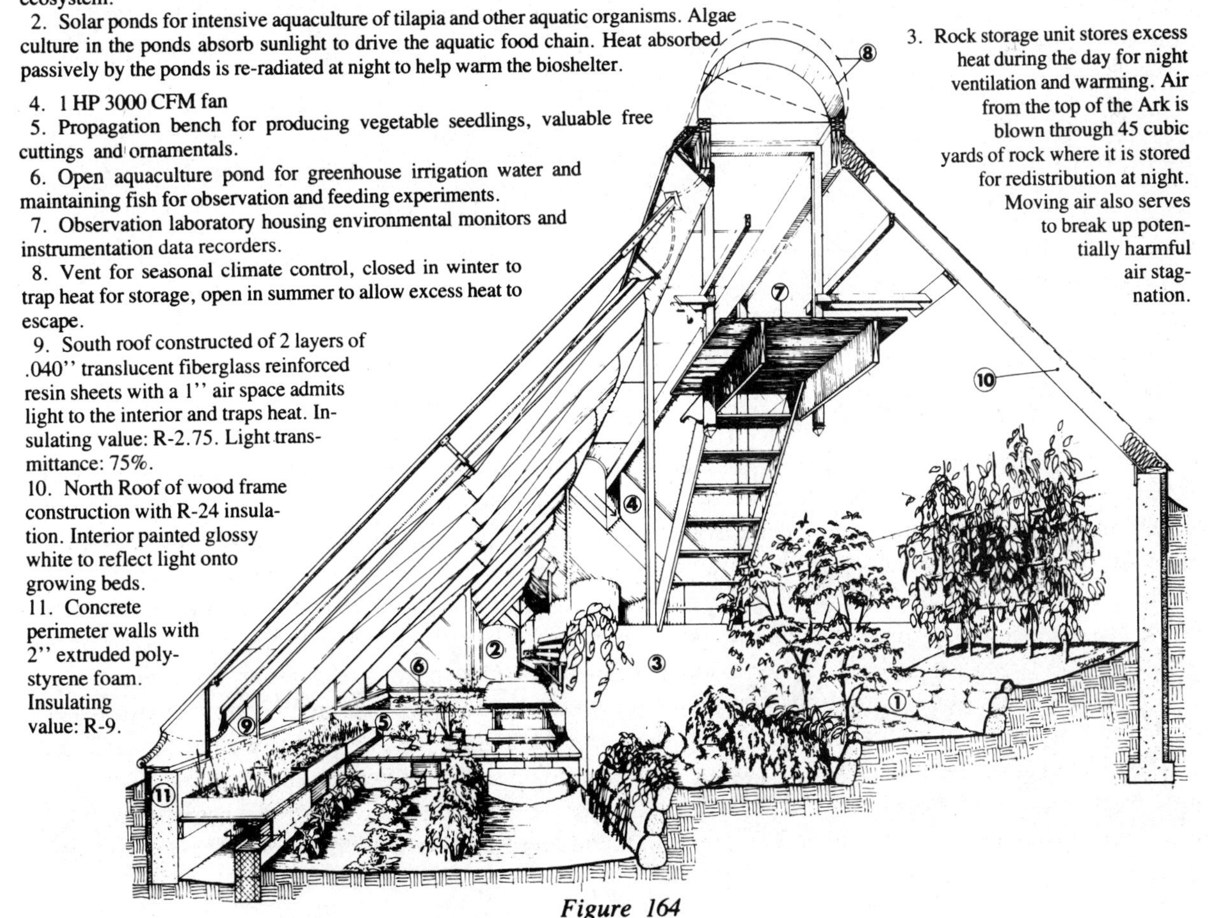

1. Terraces for year-round fruit and vegetable production employing microclimates of heat and light to nurture a diverse food-producing ecosystem.
2. Solar ponds for intensive aquaculture of tilapia and other aquatic organisms. Algae culture in the ponds absorb sunlight to drive the aquatic food chain. Heat absorbed passively by the ponds is re-radiated at night to help warm the bioshelter.
3. Rock storage unit stores excess heat during the day for night ventilation and warming. Air from the top of the Ark is blown through 45 cubic yards of rock where it is stored for redistribution at night. Moving air also serves to break up potentially harmful air stagnation.
4. 1 HP 3000 CFM fan
5. Propagation bench for producing vegetable seedlings, valuable free cuttings and ornamentals.
6. Open aquaculture pond for greenhouse irrigation water and maintaining fish for observation and feeding experiments.
7. Observation laboratory housing environmental monitors and instrumentation data recorders.
8. Vent for seasonal climate control, closed in winter to trap heat for storage, open in summer to allow excess heat to escape.
9. South roof constructed of 2 layers of .040" translucent fiberglass reinforced resin sheets with a 1" air space admits light to the interior and traps heat. Insulating value: R-2.75. Light transmittance: 75%.
10. North Roof of wood frame construction with R-24 insulation. Interior painted glossy white to reflect light onto growing beds.
11. Concrete perimeter walls with 2" extruded polystyrene foam. Insulating value: R-9.

Figure 164

For over a decade the New Alchemy Institute has been a leader in research and application of ecologically sound methods of food production and shelter. They have consistently maintained a standard of excellence in their scientific research and the quality of their publications. The Ark at Woods Hole is a working laboratory that also functions as an educational facility for community outreach programs. In it, experiments are conducted in a real world setting and open to scrutiny of the visiting public. The Institute is supported entirely by private contributions and the revenue generated by sales of publications. You can subscribe to the Institute by writing them at Box 432, Woods Hole, Mass. 02543. Please enclose a S.A.S.E.

Design and construction management for the New Alchemists is done by Solsearch of Cambridge, Mass.

Figure 165

Ecotope Group, 2332 East Madison, Seattle, Washington

This multi-disciplinary group of individuals has pioneered hands-on workshops in solar greenhouses, water heaters, and outreach education of solar technology. They have developed a fine group of workshop manuals (see Bibliography) as a result of extensive experience in this field.

The greenhouse pictured here was one of their original projects at Pragtree Farm. Over the years it has served as a working laboratory to examine both the thermal concepts in the structure and aquaculture growing within. Fig. 166 shows an interior view of the greenhouse with the reflective parabolic wall to the right. This wall acts to reflect low angle winter sunlight down to a fish tank below (Fig 167). The relatively small surface area of the top of the tank (for the volume of water) can be insulated by either a

Figure 166

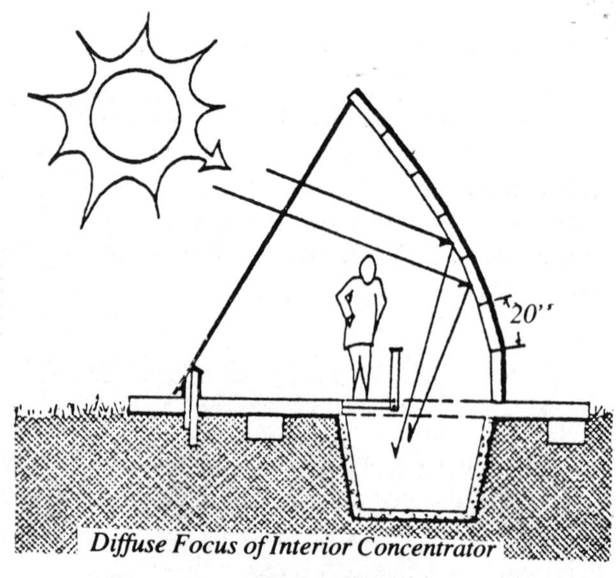

Figure 167

clear insulating cover, such as Bubble Pak, or movable floating styrofoam insulation. The tank is designed to provide both thermal storage for the greenhouse interior and a suitable environment for the fish. Technical evaluations of the concept by Reichmuth and Straub are found in Solar Energy Society conference proceedings.

Another feature of the Pragtree greenhouse is the thermal chimney shown in Fig. 165. This functions like the example given in Appendix H (Vent Sizing) to help cool the greenhouse.

David J. MacKinnon, University of Northern Arizona, Flagstaff, Arizona

Dave MacKinnon is well known in solar energy circles for his technical papers on light physics in the greenhouse and his field work and writings for Rodale Press. This small unit in Flagstaff has been the working laboratory for much of his efforts over the last five years. Here are some excerpts from his recent papers:

1. *The Greenhouse as a Solar Collector*

"As a result of this study, certain conclusions can be reached on the thermal characteristics of the Rodale solar greenhouse in Flagstaff and solar greenhouses in general.

During the monitoring period reported here, the greenhouse air temperatures never dropped below 40°F and remained 30°F to 40°F above ambient in winter and 20°F to 30° in spring.

A very rough energy analysis performed on two clear days, one in winter and the other in spring, showed that 34% of all solar irradiation entering the greenhouse was absorbed by the growing beds, the floor and the water walls.

Even though the greenhouse seemed to collect and store adequate heat in the midle of winter, some excess heat was always vented on a clear day.

Figure 168

2. *The Greenhouse as a Microclimate for Plants*

Temperatures at plant level stayed between 50°F and 85° for the majority of the winter and spring. However, different microclimates within the greenhouse itself were recognized from our study. Soil temperatures are much colder near the south glazing in winter. Although interior air temperatures increased with height during the day, little stratification occurred during the night and cloudy periods. Apparently the radiant heating from the water storage maintained a surprisingly uniform temperature environment.

3. *The Thermal Contribution of East and West Wall Storage and Floor*

Water storage along the side walls and earth at floor level provide an important addition to the thermal capacity of the greenhouse. A rough calculation shows that the sidewalls and the floor (including growing beds) each provided roughly 25% of the total stored energy.

Horizontal insulation beneath the greenhouse soil is probably not worth the effort or expense. Perimeter insulation is still recommended.

4. *Proper Placement of thermal Storage*

Our study indicates the importance of proper placement of storage. By stacking 5 gallon honey tins on the east, west and north walls surrounding the plant beds, there exists a significant radiant exchange from storage to the plant environment, dampening fluctuations and stratification in air temperatures during cold winter nights.

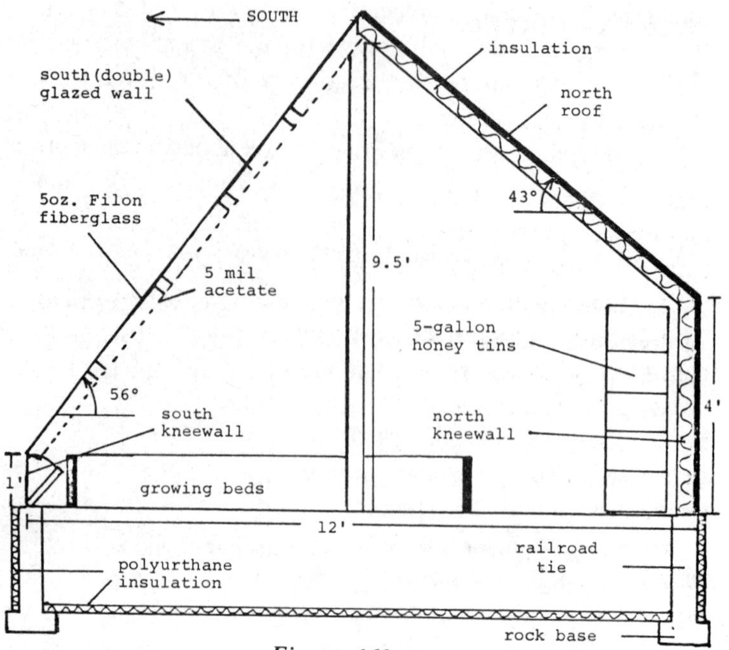

Figure 169

5. *The Use of 5 Gallon Honey Tins as Thermal Storage*

Relatively small water filled containers in a tight fitting wall seem to make a very effective passive storage system. These containers are small enough to reduce the vertical stratification of heat common in larger storage vessels, such as 55 gallon drums, but large enough to allow convection to rapidly place absorbed heat into the interior of the wall.

6. *Shading of the Greenhouse by East and West Walls*

Shading of the greenhouse interior by the east and west walls and north roof becomes increasingly evident as the spring equinox approaches. While greenhouse light levels remain quite adequate for heating the greenhouse, loss of light to the plants in morning and afternoon limits plant photosynthesis. One solution to this problem might be the use of removable insulating panels on the east and west walls that can be taken off in spring to increase the light to the plant beds (Fig. 26, p. 33). This approach for summer growing would require removal of insulating panels on the north roof as well.

Another paper examines the effects on light of the opaque north wall of a solar greenhouse:

Under almost all conceivable lighting conditions the specular north wall (shiny surface) is better than a diffusing north wall (white paint). However, snow reflection under cloudy conditions brings the light flux values at the ground from a diffuse wall nearly to those of the specular wall.

It is also true that an opaque north wall, even if reflecting, cuts out diffuse light totals in the greenhouse, but the amount is far less than one might expect, particularly if a specular surface is employed. Not to use some sort of insulated and reflecting north wall makes little sense no matter what the external lighting environment may be.

Final Note

Long wave, infrared light energy will interact with the diffuse and specular wall in a way similar to visible light energy. Therefore energy from radiant sources will be reflected from a specular wall more effectively than a diffuse wall. If this source is located at the ground or along the north wall, the specular north wall will reflect more energy to the south glazing and out the greenhouse than the diffuse wall. The presence of the specular diffuser and glazings opaque to long wave infrared will help reduce this loss during the day. But even more importantly when specular walls are used, the potential for this loss points to the need of insulation or a radiation block on glazed walls during the night.

MacKinnon's technical papers are found in the Passive Conference Proceedings and the Marlboro and Plymouth Solar Greenhouse Proceedings.

Wesley and Frances Tyson

Figure 170

I designed and built the Tyson's solar greenhouse for them shortly after they moved from New Jersey to Santa Fe. After examining their property, I told the Tysons that there was no way to attach the unit to their home without demolishing several trees. Frances didn't really care, as she considers most of the large imported trees surrounding her home water mongers anyway. But I didn't want to be the assasin. So, we decided to build an independent greenhouse on a hillside near the home. At that location the unit could be sunken about 4 feet into the hillside and the north wall bermed (earth piled against the wall above grade. See Fig. 29, p. 35.

This unit has proven to be the best in a long series of my designs and contains several important innovations that seem to work well together. The tilted back wall (after the Brace Institute design) guarantees that the plants get sufficient north light (reflected) all year. The wall is a tight sandwich of insulating materials. From the inside out it contains: ¼'' sheetrock, textured and painted glossy white; aluminum foil; 6'' fiberglass insulation; ¾'' Celotex siding; 6-millimeter black polyethylene; and 1'' overlapping rough lumber (Fig. 170).

The south wall is at a 75-degree angle to the horizon and is framed up on a low (16'') concrete-filled, pumice block wall. The southern skin is approximately 110 square feet. The south-sloping roof has alternating clear (fiber acrylic) and opaque (tin, insulation, sheetrock) panels. A roof vent is set into one of the solid panels and is opened by an automatic heat piston.

The roof may be the key to the successful performance of the structure. My brother-in-law, Paul Bunker, collaborated with me on the design. Because alternate sections of the roof are insulated, the unit stays warmer in the winter and much cooler in the summer. The solid roof sections are also perfect for mounting hinged movable panels if extra insulation were needed. In this greenhouse there is only 50 square feet of clear roof approximately normal to the summer sun.

Another important factor in the design of the Tyson greenhouse is that it is oriented 25 degrees southeast of due south. This angle was determined mostly by the lay of the land and our desire to alter the natural environment as little as possible.

Figure 171

Since very few morning rays would enter the greenhouse through the east wall at this orientation, the east wall is 70 percent opaque and insulated. The inside of this solid wall is reflective (tomatoes that are situated directly in front of it show excellent blossoming and fruit set). The west wall is clear.

This greenhouse was designed to contain ten water drums and have movable insulating panels on the roof. With these provisions, I estimated that the unit would maintin 50 degree interior temperatures at -10 degrees outdoors. With only four drums installed and without the styrofoam panels in place, the greenhouse held a 43 degree low at -4°F outside. That's truly remarkable, and I can only attribute it to the roof design and the mass of the insulated block wall (it has an inch of styrofoam on the outside surface).

Frances Tyson is an experienced gardener. For years she had a large organic garden back in New Jersey. Frances uses growing techniques that would make a chemically oriented gardener turn over in his sodium nitrate. She uses relatively "hot" manure, heavy green mulch, never sterilizes anything, and plants extremely densely. I've never seen a greenhouse that could match the level of production she gets out of the space she has. Some of Frances' techniques are discussed in Chapter VII. This amazing woman also finds time to build solar collectors out of beer cans, and to speak to anyone who will listen about the development of solar energy and the abolition of nuclear power.

Owners: Wesley and Frances Tyson
Designer: Bill Yanda
Builders: Paul Bunker, Susan and Bill Yanda
Floor Area: 180 square feet
Clear Area: South face = 110 sq. ft.; south roof = 50 sq. ft.; east wall = 25 sq. ft.; west wall = 50 sq. ft.
Main Function: Vegetable production for personal use.

Thermal Storage: Pumice block walls filled with concrete = aproximately 7000 pounds; 220 gallons of water (enclosed drums)

Material Costs: $5.00 per square foot (1977)

Barbara and Peter Voute

The Voute's greenhouse is the geometry of many of the community structures found on pp. 152-3. To my knowledge, this was the first design of its type built in the U.S., and it's a good one *if* provision for north lighting is included in the roof. Actually, the design which is very old is one of a large number of geometries studied by J.C. Loudon in *Remarks on the Construction of Hothouses* published in 1817.

The south face is at a 65 degree tilt, and the clear roof section slopes away from the front (toward the north). The two clear panels running the length of the roof are used only in the summer for north lighting. In the winter they are covered with rigid styrofoam to reduce heat loss.

In designing this unit, I experimented with a glazed direct-gain wall. I think it's a good idea that needs further development. This is how it works.

The coldest area in the greenhouse is in the extreme south bed at ground level. I knew that if the earth there could be heated, the plants would do better through the long, cold, winter nights. The south face is framed up on a 16'' high by 10'' thick concrete wall. I painted the outside of this low wall dark brown and glazed it with one layer of fiberglass acrylic material. It was well sealed on the top and bottom. Heat is transferred through the concrete to the earth bed directly behind it. (This is like a Trombe wall but has no vents for air circulation.)

The results were that the young plants nearest the wall showed faster, healthier growth than those farther back in the bed. However, the stored heat wasn't enough to carry some of them through the coldest winter nights. When the outside temperatures dipped to -7°F, a patch of beans in the front bed froze. Nothing else was damaged, including tomato plants situated centrally in the unit. But Dr. Voute was disappointed and he bought a small electric space heater calibrated to turn on at 45 degrees.

Figure 172

To improve the performance of the direct-gain wall I would: 1) make it thinner; 5'' would be thick enough for strength and would provide a higher rate of heat transfer to the front bed; 2) double glaze the south low wall; this would prevent such rapid conduction losses back through the wall at night; and 3) insulate the lower clear wall (at night).

The Voute's grow some beautiful flowers and vegetables in this greenhouse. The soil mixture they use is 1/3 city sludge from the sewage plant, so occasionally they'll have a healthy tomato or chili plant pop up in an unexpected place. Dr. Voute has also developed a rather intense personal relationship to his unit. To paraphrase him: ''The damn thing's like a spoiled pet or child; it needs attention all the time. I think we'll skip December and January growing next year and take a vacation from it.'' This is said lovingly, believe it or not.

Owners: Peter and Barbara Voute
Designer: Bill Yanda
Builders: Paul and James Bunker, Bill and Susan Yanda
Floor Area: 160 square feet
Clear Area: 128 sq. ft. (south face); 40 sq. ft. ft. (east and west walls); 32 sq. ft. (roof)
Thermal Storage: 3 cubic yards of concrete in walls (approximately 12,000 pounds) and 330 gallons of water
Supplementary Heat: Electric space heater
Material Costs: $4.00 per square foot (1976)
Main Function: Vegetable and flower production for home use.

Community Solar Greenhouses

USDA Solar Outreach Program

Figure 173

There is a new sense of community building emerging across the United States with solar greenhouses the crystallizing nucleus. A greenhouse building workshop, such as the one above, held in Pittsboro, North Carolina, brings citizens from diverse backgrounds together for a common goal: learning to design and build a solar greenhouse. Now there are a large number of trained solar crews out working in their communities building and teaching. You can be a part of this educational experience. It is the best possible way to learn the technical specifications and practical techniques involved in building a solar application. We've included a listing of 'hands-on' solar training groups in the Bibliography, p. 201.

These community greenhouses, though small by commercial standards, are helping to create a consciousness of food independence and self reliance, often within the urban setting. In most cases, the structures were built by previously unemployed youth or poverty victims, and skill training is one of their primary goals. Another goal for these projects is economic self-sufficiency; produce from the unit must support its management upkeep. These structures represent a bold step in the quest for energy independence and meaningful employment in the solar future.

Figure 174

Colorado Rocky Mountain School, Carbondale, Colorado

This two-zone greenhouse was designed by Ron Shore (now of Thermal Technology Corporation) and built by students as an inter-disciplinary project at school. It incorporates 350 sq. ft. of south glass and 1575 gallons of water storage. A 1/3 horsepower fan pushes 1,500 cu. ft. per minute of air through 129,000 pounds of rock storage. The effects of this mass and rock storage is evident in the fact that the coldest temperature yet recorded in the structure was 38°F at -29°F.

Figure 176

La Jara, Community Action Program, Arnie and Marie Valdez

This greenhouse is one of several similar designs built by the industrious Valdez family. Arnie and Marie, from San Luis Colorado, are well known in the western U.S. for their work with migrant workers and the National Farmworkers Organization. They conduct design and building workshops for water heaters, crop dryers and flat plate collectors, as well as greenhouses. This greenhouse is 16' wide by 36' long and has a 65° south glazing. Notice the airlock, which also serves as tool storage on the east side.

Figure 175

Citizens for Citizens Project Fall River, Mass.

Designed and built by Mike Hurly. One of the main goals is to involve urban families of the area in year round gardening. The 1000 sq. ft. structure contains 3000 gallons of water in drums. Coldest temperature in '77-'78 was 38° at 0° outdoors. Normal winter lows were the 45°–50°F range. Heavy emphasis is on spring seedling production to complement the urban gardening program.

Montachusetts Opportunity Council, Leominster, Mass.

This project headed up by Dennis Jaehne serves two groups. Land for the greenhouse was donated by the local school system, and students receive training there. Also, senior citizens have individual garden plots within the greenhouse that they manage.

Figure 178

Figure 177

Jaehne has written a wonderful paper on the frustrations, accomplishments, and promise of this project. It is published in the 3rd National Passive Solar Conference Proceedings and is mandatory reading for any person or group considering such an undertaking.

The National Center for Appropriate Technology, Butte, Montana

The National Center for Appropriate Technology has actively supported solar greenhouse research and implementation. Recently, they have begun a national monitoring and evaluation program that should supply a great deal of needed information. This 16' x 40' greenhouse is a research facility in Butte. There are two sections to the greenhouse separated by the vestibule entrance in the center. One experiment involves four 55-gallon drums plumbed in series and mounted directly over the entrance. The incoming water is at 43°F. The black drums act to preheat the water to $70^{o}-80^{o}$F during the spring-fall period. The warmer water is used for crop irrigation.

Figure 179

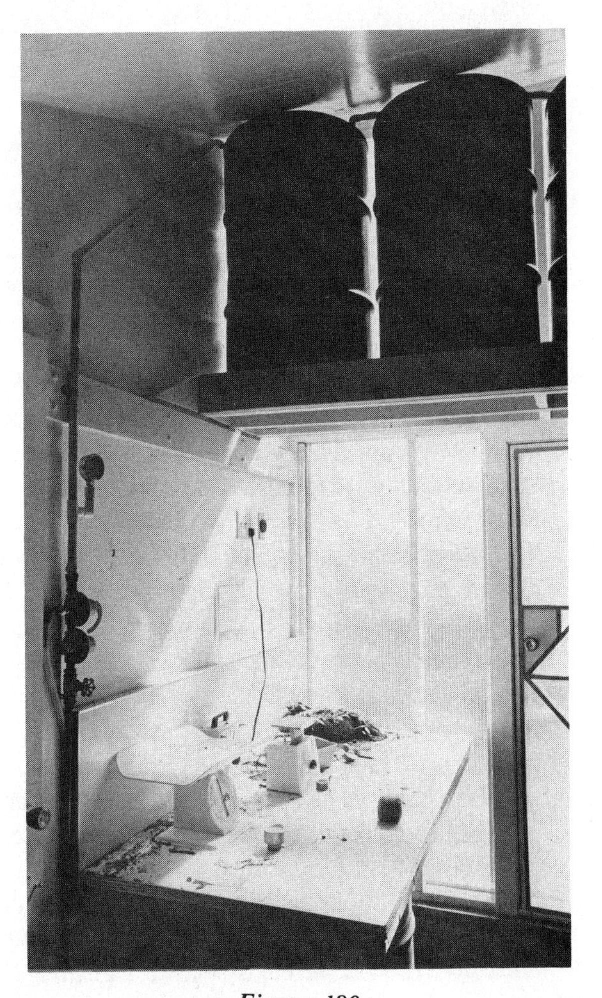

Figure 180

Solar Greenhouse Manufacturers

This section reviews commercial manufacturers of thermally efficient greenhouses (and units that can be easily modified to become true solar greenhouses). We will also mention producers of innovative and important accessories such as movable insulation and thermal storage systems.

The Vegetable Factory, Long Island, New York

Vegetable Factory offers a complete multiple-purpose structure consisting of framework, panels and accessories. It can be utilized as a solar-panel greenhouse for growing; or serve as a highly efficient passive solar collector for heating. To operate the unit as a passive solar collector, a heat-storage system should be added.

Figure 181

On Nantucket Island, off Massachusetts, Wade Green uses a Vegetable Factory solar-panel, lean-to greenhouse as a passive collector attached to a Trombe wall. The concrete-filled cinderblock wall, constructed with air vents at top and bottom for natural convection air currents, is painted black to assure the highest degree of heat absorption. At night, an aluminized mylar curtain is drawn in front of the wall to keep the collected air and heat inside. During the daytime, the curtain is removed to once again allow the solar energy to collect and warm the house (Fig. 182).

In Mr. Green's published *New York Times* article of March 9, 1978, he states: "At the end of a sunny day, the wall is nearly as hot as a steam radiator, and slowly gives off its warmth throughout the night." Mr. Green also confirms that with his passive system, even without the benefit of extra-storage capacity, heating-fuel costs are reduced 40 to 50 percent.

Figure 182

Vegetable Factory also offers standard-size and custom panels, without the framework of the house, for do-it-yourself solar collector projects.

In many applications, complete greenhouses, or special panel construction has been substituted for roof tops. Fig. 183 shows the "solar office" of a New York doctor. In this application, the solar panels help heat the office, creating both a conducive growing room for plants and flowers, and a pleasant consulting room for patients.

Many schools and institutions are now utilizing solar-panel Vegetable Factories as "greenhouse classrooms." As an energy saver, the greenhouse collects solar energy which supplies additional heat to a school, and at the same time provides the proper atmosphere for plant and vegetable gardening as a real learning experience.

The shatterproof quality of both the acrylic-fiberglass and GE Lexan panels is an important safety factor where children are concerned.

For complete literature and pricing information, write Vegetable Factory, Inc., Dept. SG, 100 Court Street, Copiague, Long Island, New York 11726.

Figure 183

Solar Technology Corporation, Denver, Colorado

Solar Technology Corporation (SOLTEC) designs, manufacturers, and markets solar homes, building systems, and solar products for residential, commercial and industrial applications. Over five years of intensive solar engineering and product development has produced a diversified line of products. Their on-going product development program is directed towards providing innovative solar products for the continuously expanding domestic and international solar market.

SOLTEC is currently manufacturing and making its SUNUP^{tm} solar building system and the SOLAR GARDEN^{tm} greenhouse.

- SUNUP^{tm} collectors employ air as the heat transfer fluid and have a variety of uses for *space and water heating*. Collectors may be mounted on south roofs or walls of most buildings, or on ground-mounted solar furnaces. *Solar sauna* and *pool heating* applications using the SUNUP^{tm} collectors are possible.

- The SOLERA^{tm} solar building system is a prefabricated, modular, expandable "walk-in" collector that provides clean solar heat to homes while serving as a *solarium, home addition,* or *pool enclosure*. SUNUP collectors may also be used in SOLERA to increase performance and convert it to *sauna* and *water heating* uses.

- SOLAR GARDEN components transform SOLERA into a *solar greenhouse,* which provides year-round solar heated growing space for flowers, fresh fruits and vegetables, as well as heated, humidified, oxygen-rich air to attached homes (Fig. 184).

Figure 184

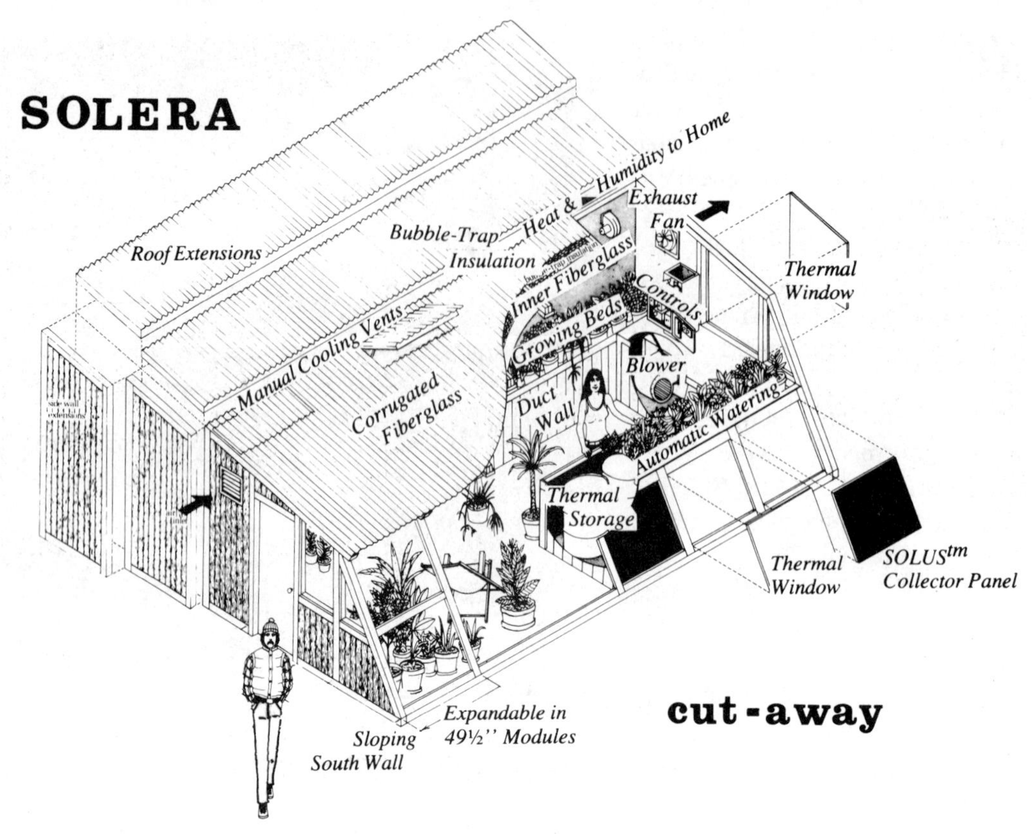

Figure 185

The mult-functional SOLERA may be easily and inexpensively surface-mounted to new or existing homes in freestanding, attached or abutted positions. With a modest amount of building experience anyone can install this adaptable solar energy system.

The SOLERA low-temperature solar collection system traps the incoming solar radiation with highly insulated wall panels and south-facing thermo-windows. Optional inner windows, pop-in panels and hot-air collectors reduce heat losses at night and during cold weather.

Solar heat is stored passively by direct radiation absorption on the black-painted surface of thermal storage tanks. Thermal transfer and storage efficiency is improved by using a combination blower, air handler and duct network.

Heat is drawn into attached homes with fans or by natural air circulation through open windows or door. A conventional hot-air heating system may also be used to distribute solar heat within the home.

Cooling of the unit and attached homes is achieved with many optional systems including manual roof vents, exhaust fans, nighttime cool storage, and evaporative cooling. Passive heating and cooling techniques enable the SOLERA to operate without electricity, although thermostatically controlled fans allow greater temperature control.

Solar Technology Corporation (SOLTEC) has an on-going research and product development program directed towards advancing a diversified product line including commercial solar greenhouses and higher temperature solar systems. They also offer solar consulting, design and custom building services.

Four Seasons Solar Greenhouses, 672 Sunrise Highway, West Babylon, New York

Four Seasons is a company offering a wide range of features in solar greenhouses. They are one of the few manufacturers with a triple glazing option of GE Lexan, plus a triple glass system. They also sell hot water collectors and air exchange systems custom designed for their product.

Figure 186

Solar Room, Solar Resources, Inc., P.O. Box 184, Taos, New Mexico

Figure 187

Since the first edition of this book, the Solar Room has proven itself to be one of the best buys in "BTUs for the buck". Literally millions of American homes could save heating dollars immediately by the installation of a Solar Room. Every time a leading national politician or official gets on television to say that the nation is "still waiting for cost effective solar heating," the president of Solar Resources, Steve Kenin, fires off about a dozen irate letters telling them where to go visit a satisfied Solar Room customer to find out the truth.

With or without national political endorsement, the Solar Room continues to make more and more solar converts. The product comes in kit form and can be erected by a couple of novices in 3-5 hours. Basically, the Solar Room is an air

Figure 188

tension structure which gives it a tremendous amount of strength for its very light weight. The double skin is one continuous piece, so air infiltration thrugh cracks or joints is practically non-existent. Besides the greenhouse model shown here, the company also manufacturers and distributes a system that only extends a few inches and transforms the covered south wall into a solar collector.

Here is Steve Kenin's explanation of his product:

"The Solar Room is a device that turns the southern side of a home into a solar heater. Made of a special plastic, a Solar Room can supply 35 to 65% of home space heating needs. With heat storage and insulation options, its heating capacity is greatly increased. The Solar Room is available in kit form and is designed to be an exterior room, seven feet wide and as long as space permits; 20, 30 or 40 feet. The longer the Solar Room, the more heat is collected.

"Not only a heat collector, the Solar Room is a versatile, inexpensive addition to the home, costing $3.50 to $4.50 per square foot of floor space. It is an airtight, thermally efficient space, and can serve as a greenhouse, a winter playroom for children or as a foyer to the house where coats, boots and bicycles can be stored out of the winter weather. As a greenhouse, the Solar Room is an especially efficient space, providing warmth for the household and fresh vegetables for the dinner table. In the spring the garden can be started early in the greenhouse and transplanted outside when danger of frost is past.

"The Solar Room is also a 'take-down' room. Because of its effectiveness as a heat collector it is not needed during warm weather, and has been designed to take down during the summer months. The initial installation requires less than a day's time, and after that removing the Solar Room in the spring and putting it back again in the fall takes only a few hours. When not in use, it takes up little storage space.

"The Solar Room kit is made possible by the use of a special plastic that resists the disintegrating rays of the sun and lasts for years, special aluminum extrusions that hold the plastic in place, the best grade of clear heart redwood that will not rot in contact with the ground or in moisture, and galvanized ribs that support the plastic skin and will not rust. The Solar Room is double-glazed, which means there are two layers of plastic with an insulating 'dead-air' space in between. It has withstood winds over 60 miles per hour."

I will add that the air supported, slippery glazing also sheds heavy snow loads quite effectively (as proven by Wisconsin and Michigan customers). The portability of the Solar Room makes it particularly attractive to renters who desire a greenhouse and pay their own utilities. Anyone serious about saving money on heating bills who does not have the time or capital for custom building should investigate the Solar Room.

Helion, Inc., Brownsville, California

Mobile homes are prime candidates for solar greenhouse additions. If the mobile home has its long axis oriented east-west, a large area of southern wall will be available for the greenhouse/collector. Because of the narrow width of the home, air circulation and communication to the living area will be optimum. A major portion of winter heating needs will be met by the greenhouse alone.

Jack Park of Helion, Inc., has designed many solar greenhouses and sells blueprints for a mobile home retrofit. Helion also produces a microcomputer for home and agricultural energy management and monitoring.

Jack Park, Helion, Inc., Box 455, Brownsville, California

Figure 189

The following manufacturers market movable insulation systems that greatly improve the thermal efficiency of solar greenhouses and other solar applications.

Appropriate Technology Corporation
Brattleboro, Vermont

The Window Quilt Insulating Shade R increases the R-value of double glazing up to 5.5 when installed. It is made from a polyester fabric quilted for extra strength (Fig. 190). The ATC quilt can operate at angles up to 45° and includes a track system at both edges to insure a complete seal. For information on distributors in your area contact: Appropriate Technology Corp., P.O. Box 975, Brattleboro, Vermont 05301.

Figure 190

Thermal Technology Corporation
Snowmass, Colorado

The Curtain Wall consists of a number of layers of thin, reflective material rolled down by motor to cover large collecting surfaces. Radiation from the structure's thermal mass (in winter mode) or from the sun (in the summer mode) is intercepted by the curtain and warms the air between the layers. This causes it to occupy more volume, inflating the curtain as air expands within its multilayered envelope (Fig. 191). Storage area and power consumption for the automated curtain is minimal. A curtain 24-feet long x 16-feet high can be stored in a 6-inch diameter roll. For details and thermal performance data, contact: Thermal Technology Corporation, P.O. Box 130, Snowmass, Colorado 81654, (303) 963-3185.

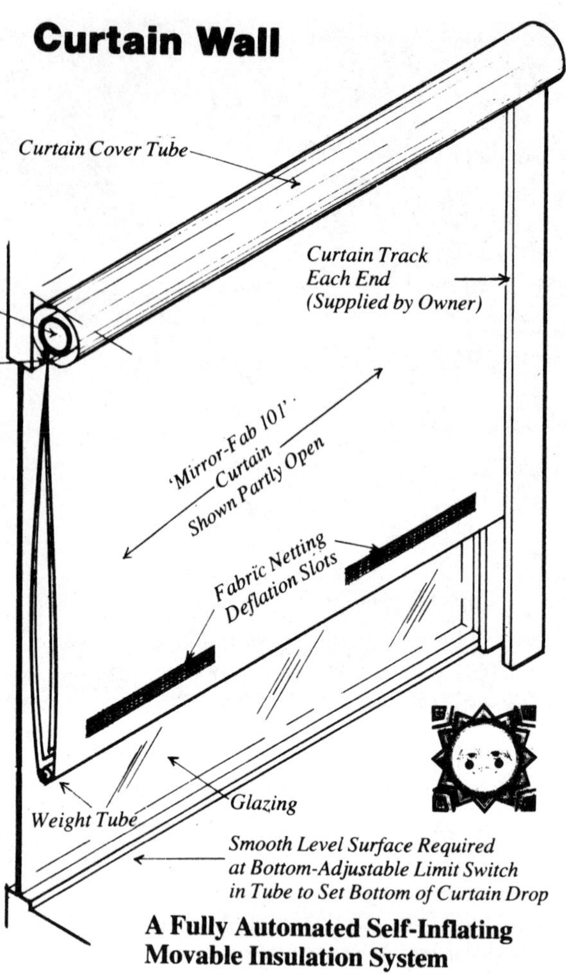

A Fully Automated Self-Inflating Movable Insulation System

Figure 191

Is Company, Guilford, Connecticut

This insulating window shade is applicable to vertical glazings and may be purchased in sizes up to a maximum width of 66 inches and length of 96 inches. The shade is made up of five layers of plastic film that separate when rolled down to form a series of dead air spaces (Fig. 192). The interior layers of the shade are made of metalized film to cut radiant losses. The IS high "R" shade in combination with double glazing has been tested to produce an R-15 barrier. Extruded plastic head and jamb frames (Fig. 193) contain the shade. If desired, the builder may construct wooden enclosures (from IS Company plans) to meet specific needs.

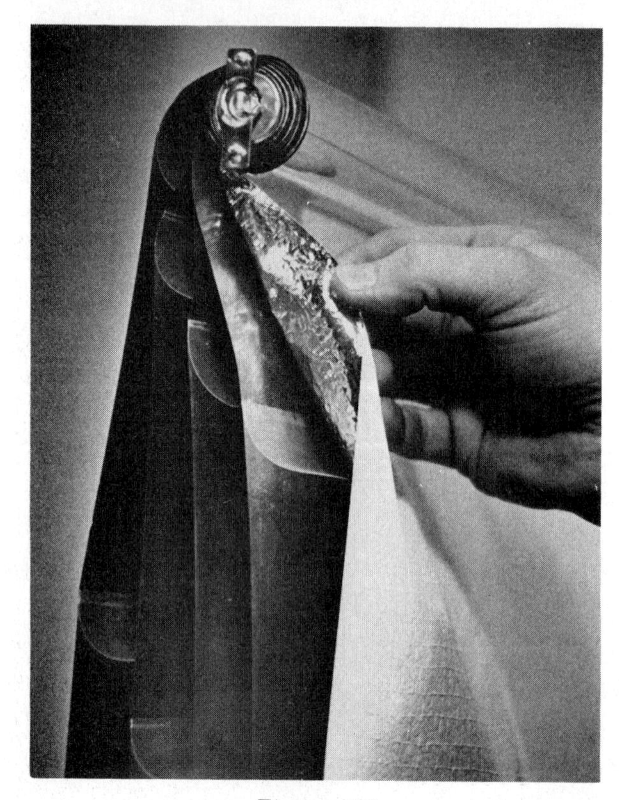

Figure 192

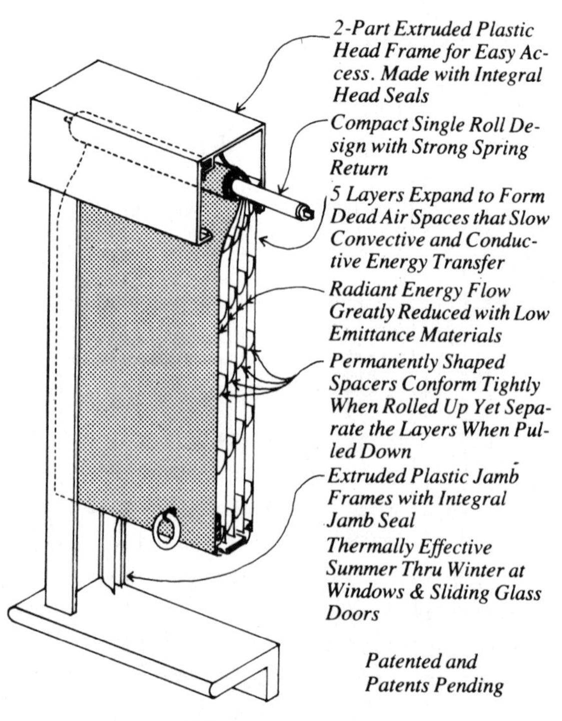

Figure 193

Zomeworks, Albuquerque, New Mexico

Zomeworks and its president, Steve Baer, need no introduction as innovators in the field of solar energy. Three of this company's developments have important implications to solar greenhouses.

Figure 194

The Beadwall^R system seen in the Monte Vista greenhouse was completed in 1973 (see Fig. 194). It was invented by David Harrison of Zomeworks and solves the tricky problem of how to combine movable insulation and light-transmitting clear surfaces. In this system, insulating styrofoam beads are blown in between the two panes of glazing (3'' apart) when temperatures drop below tolerable levels. In the morning, or when temperatures rise, a photocell switch tells the motors to evacuate the cavity, and the beads are sucked back into their storage containers. I've seen Dave get a standing ovation from audiences at solar conferences when he was demonstrating this device. The Monte Vista greenhouse was one of the first applications of Beadwall^R and used six vacuum cleaner motors and storage bins to accommodate the beads. Besides greenhouse applications, this system can be retrofitted or newly installed in home windows to maximize direct gain and almost eliminate heat losses.

Nightwall^R is a poor man's Beadwall^R in which *you* supply the motive power to move panels. They are simply rigid styrene or styrofoam sheets that are cut to fit the exact dimensions of window or greenhouse clear wall. Zomeworks supplies magnetic strips with sticky backs, small metal contacts to attract the magnets, and an instruction sheet to help you put it together. When the magnets are attached to the perimeter of the window and the metal strips are adhered to the styrofoam panel, the magnetic force will tightly bond the insulation to the window at night. You remove the panels in the morning for transmission of light and direct gain. As the clear areas in a greenhouse are the major source of heat loss in the structure, they are prime candidates for Nightwall^R installation. The rigid panels could also do double duty as reflectors behind plant beds during the day. Nightwall^R will work well on a vertical or almost vertical surface in a greenhouse. Zomeworks also supplies a chart to show you how many BTU's the Nightwall^R can save on standard sized windows in various parts of the country. It's amazing what these simple applications can do for you.

Figure 195

A third Zomeworks product with greenhouse potential is Skylid^R. It's a device that uses counter-balanced weights and freon containers to open high vents or skylights. Steve uses Skylid in his home and greenhouse (Fig. 195). Six Skylids operate 240 square feet of insulated louvres to prevent heat loss. Baer's greenhouse also has six 30-gallon drums for thermal storage and 30 tons of rocks. This greenhouse is a single-glazed structure and has held 32-degree minimums in 5 degree outdoor lows. Baer is also using Skylid^R mechanisms as passive trackers for concentrating collectors.

These active and passive improvements provide important options for the solar greenhouse owner. As we noted in Chapter V, the beginner might consider the simpler, passive approach first. If you do want one of the more complex active systems, Zomeworks is definitely the place to go for it. They also sell plans for "bread box" water heaters. For more information, write to: Zomeworks, P.O. Box 712, Albuquerque, New Mexico 87103.

One Design, Winchester, Virginia

Traditionally, oil drums full of water have been the simplest form of water containment for solar greenhouses. Now there is a new, engineered, mass produced water module that can be stacked high with negligible pressure, buried in crushed stone as greenhouse floor, placed under planting tables, or anywhere within the structure.

The modules, by One Design, Inc., of Winchester, Virginia, are made of fiberglass and were specifically designed to offer placement *versatility* of application, to eliminate the high pressure associated with vertical tubes, and to offer higher mass by using rectilinear shape. Toward these goals five generations of waterwall modules were developed and tested in homes in northern Virginia and Pennsylvania. The result of these demonstration/test houses is a glass-fiber module suitable for widespread utilization.

The modules are made of glass reinforced polyester, one tenth of an inch thick. The black, absorbing pigment, as well as an ultraviolet stabilizer, is contained in the fiberglass resin. As they are stacked, each unit receives a lid. They are then filled with water via a standard garden hose connection. No plumbing is necessary.

Figure 196

The rectilinear configuration provides 27% more mass than tubular containers in a given area. In addition, earlier vertical designs proved to be problematic due to high water pressure. The horizontal configuration reduces water pressure to a negligible level (less than 1 lb. per sq. in.)

Figure 196 shows four stacked modules deployed for collection (size of wall approximately 94" x 100" x 20'). The units can be concealed between an interior finish wall and the south glass. When placed between a greenhouse and living space, they can be drained of water and nested out of the way for summer storage. In this way a greenhouse area can be joined to the living area during summer months.

The modules can also be buried in crushed stone, the top of which provides the floor surface of a greenhouse. They are then filled with water and are out of the way. Modules can be stacked as intermediate and north walls when roof apertures admit sunlight. They can be stacked as room dividers and can be retro-fitted into add-on greenhouses.

Product Specifications

	Model A	Model B
Length	94 in.	94 in.
Width	16.5 in.	20.5 in.
Height	24.5 in.	24.5 in.
Weight Empty	24 lbs.	30 lbs.
Weight of Contained Water	800 lbs.	1,000 lbs.
Heat Storage Capacity at 40° Rise	31,500	40,000

	Model A	Model B
Wall Thickness	.01 in.	.01 in.
Material	Ultraviolet stabilized fiberglass	
Heat Extractor Rate	Determined by installation	
Water Pressure at Bottom of Stack	¾ lb/in^2	¾ lb/in^2

Kalwall Solar Components, Manchester, New Hampshire

Kalwall manufactures and distributes a full line of solar products and accessories. Many are quite relevant to greenhouses, as you will see if you get their catalog listed on p. 200. One line, however, warrants particular note at this point. These are the heat storage tubes that are filled with water and used in greenhouses and other solar applications. The following information on them is reprinted from p. 33 of the 1979 catalog.

Sun-Lite^{R} Storage Tubes & Containers

Sun-Lite Tubes and Tanks have been utilized by a number of research groups and colleges experi-menting in the field of aquaculture. When ordering tubes or tanks, please specify the use, such as "heat storage" or "aquaculture."

Friction fit caps and/or black coating either 180° or 360° around the tube are available on a custom basis. Please refer to the catalog price guide.

Sun-Lite^{R} Storage Tubes are a key component in Kalwall's Solar Furnace.

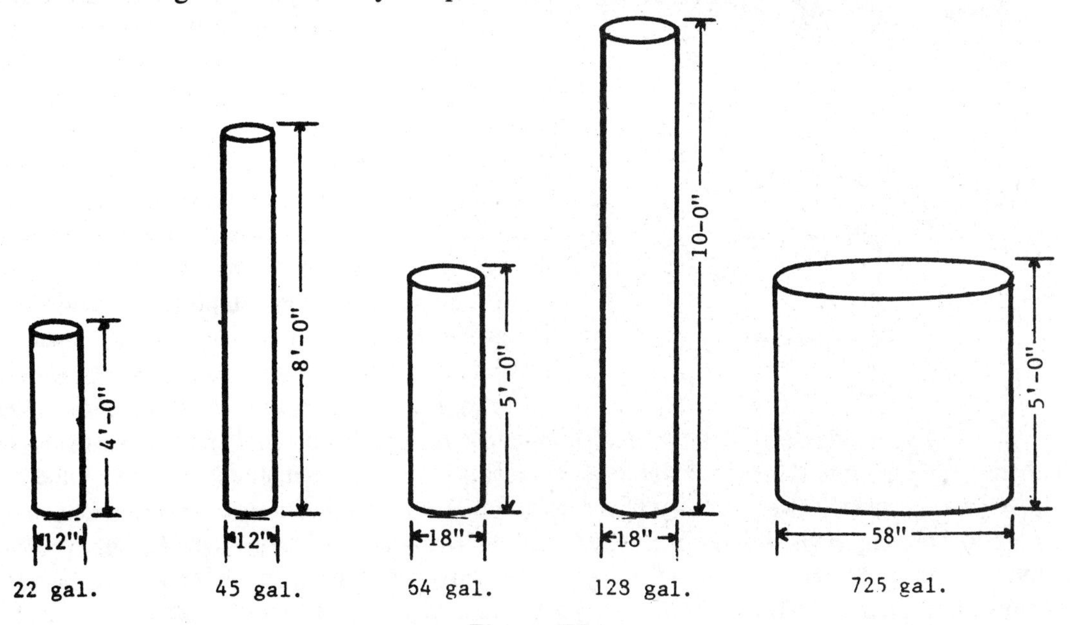

Figure 197

Average Properties

Diameter of Tube	12"	18"
Height	8 ft.	10 ft.
Volume	6.28 ft.^{3}	17.67 ft.^{3}
Wall Thickness	.040"	.040"
Material	Sun-Lite^R	Sun-Lite^R
Weight Empty	10 lbs.	20 lbs.
Weight of Contained Water	392 lbs.	1102 lbs.
Heat Storage Capacity	392 BTU/40°F.	1102 BTU/°F.
	15,630 BTU/40°F. Rise	44,080 BTU/40°F. Rise
Typical Heat Extraction Rate(1)	1,400 BTU's/Hour	2,100 BTU's/Hour

(1) Typical Heat Extraction Rate—stored heat energy can easily be extracted by passing air over the surface of the tube. This rate depends on motor, fan and ducting used. The listed extraction rate is estimated for 500 feet per minute with adequate air volume, and the tubes at 30° above air inlet temperature.

ALTERNATE GLAZING METHODS

Wood

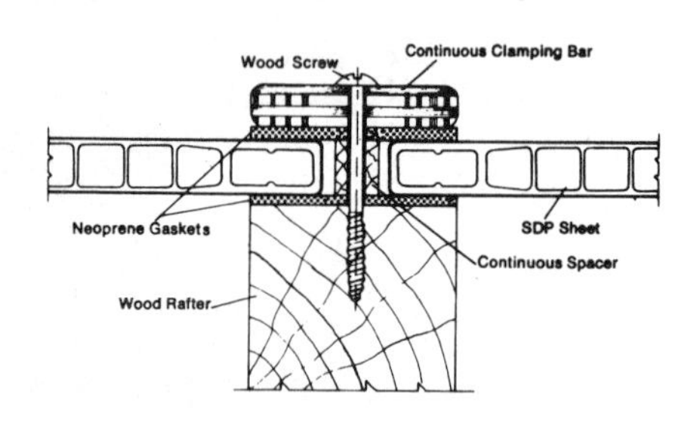

Steel / Aluminum

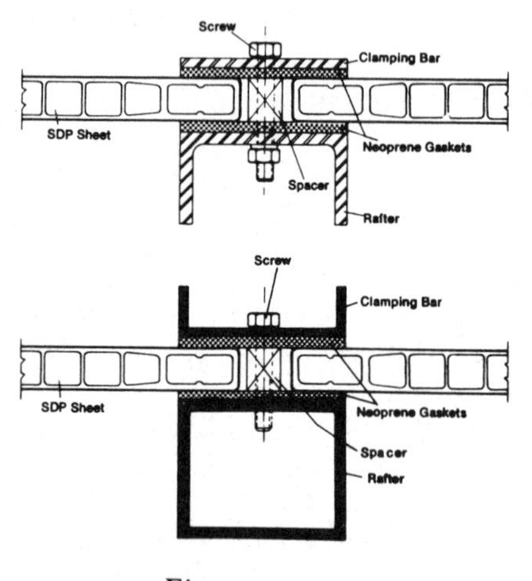

Figure 198

Cyro Acrylite SDP

This is a fairly new product on the American market and one that we've been very pleased with. The lightweight ribbed acrylic panels are attractive, high in solar transmission, and have about the same heat loss characteristics as double glass. The material diffuses the light quite well and, because it is acrylic, should have a long life. Here's a perfect roof application because of the light weight and strength of the product. It is 5/8" thick and comes in 4 x 8, 10, 12' panels. In small quantities, the manufacturer CYRO Industries of Wayne, New Jersey will refer you to their local distributors.

Heat Motor and Solar Fan

Two other products we've used extensively and can endorse are the Heat Motor and the Solar Fan.

The Heat Motor is a temperature activated piston that is mounted in the roof of the greenhouse to push open a vent or series of vents. Its opening can be regulated and set by turning a collar at the top. This gives a great deal of control. It can be set to open anywhere between 60°F–100°F or, in a winter mode, not to open at all. The ram at the top will push 50 pounds 12 inches. By proper mounting on the hatch, a full open position much larger than 12 inches can be achieved. There are other similar heat pistons available (Solar Vent and Thermofor are two good ones). However, for strength and dependability, the Heat Motor is hard to beat. I once disregarded the instructions and didn't mount a safety chain on a vent unit. A heavy wind blew the vent open and completely disemboweled the Heat Motor, pulling out the internal contents (a sticky grayish gunk) and depositing them all over a greenhouse roof. My brother-in-law, Paul Bunker, collected the innards by scraping them up with a Red Devil knife. He shoved them back into the Heat Motor and screwed down the collar. Four years later the thing just keeps on opening and closing everyday.

The Solar Fan made by the William Lamb Co. is a high quality air mover powered by a small photovoltaic array. The company sells it primarily as an attic ventilator, but it is perfect for a small greenhouse. The 12-volt D.C. unit can push over 600 cfm quietly and dependably. In the Solar Sustenance Office we have designed a mount that gives it three operating modes. In the photo it is in the exhaust mode moving air out of the greenhouse in the summer. Come fall, when heat is needed by the office, the fan is pivoted 90° to blow apex greenhouse air to the back room fifteen feet away. The third mode is a switch that directs the power down a line to charge a 12-volt battery. The attribute I enjoy most about the fan is its high quality. There's nothing second rate about it. The heavy little motor runs without a sound and must have sealed bearings because I can't find anywhere to oil it. (I don't anticipate the need, either.) The photovoltaic array is totally sealed against weathering.

Finally, a built-in feature makes the unit unique. Because it is linked directly to the sun, the Solar Fan regulates itself. The fan works the hardest when the sun is brightest. So it's going full tilt on a bright day to charge an adjoining room or exhaust outdoors. When it gets cloudy, the Solar Fan slows down or stops completely. It's like having a tiny little brain linking the power of the sun to the spinning of a fan blade—for your benefit. Both of these products demonstrate remarkable and totally beneficial use of technology.

For the owner-builder, many of these products are hard to find. Here's a checklist of things to do in order to locate them.

- Most of these manufacturers would prefer that you make small purchases through one of their regular distributors. You can write them and ask who's selling the equipment in your neighborhood.

- Look under Solar Products in the Yellow Pages. There is beginning to be quite a healthy little industry in solar stores and retail distributors. Look first for the store that sells a variety of products, not just one line they want to push down your throat. Ask them to give you their experience with the product or the names of previous customers you can contact.

- Here's the name of one distributor that carries most of the products listed here at excellent prices, and will drop ship them to you.

  Brother Sun
  Solar Greenhouse Glazings
  and Supplies
  Rt. 6, Box 10A
  Santa Fe, N.M. 87501
  (505) 471-1535

Heat Motor and Solar Fan in Operation

Figure 199

Photovoltaic Array for Solar Fan

Figure 200

Specializes in Acrylite, Lascolite, Double Glass Units, Heat Motor and Solar Fan.

- In the Bibliography (p. 200) you will find a listing of good catalogs that sell equipment mail order. You will usually pay a slightly higher price doing business this way, but it may be worth it in time savings and being able to examine a wide range of products.

APPENDIX A

YANDA'S HOME HEATING CONTRIBUTION CHART

Using the Home Heating Contribution Chart

The chart and the procedure that follows will give the heat output of the attached solar greenhouse throughout an entire heating season *as a percentage of the total heating* requirements of the home. It is designed for owner/builders, contractors, and workshop people who are planning add-on greenhouses but have neither the time nor the money for a sophisticated design analysis or access to computer models. The results of several dozen tests have been compared to monitored greenhouses in the field and computer simulation models. They have been found to be accurate to within 10% of high solar fractions and 25% of lower solar fractions. After you've used the chart a few times you'll be able to compute a predicted performance for any greenhouse in any climate in about five minutes.

A Few Words of Caution . . .

By not using common sense you can exceed reasonable performance estimates. For instance, this procedure assumes you have a good solar site with no appreciable shading to the south. You will have to reduce the solar fraction proportionally to any 9 AM–3PM winter shading at the site. Also, please don't assume that you can double the performance of a solar greenhouse by simply doubling the area of the glazing. This can be achieved at low solar fractions, but if your design is providing 60% as is, doubling the size won't get you 100% solar heating with 20% to give to neighbors.

I wouldn't build a new solar greenhome with this chart alone. You can use it to extract ideas, incorporate them into the initial design phase, and then take your rough concept sketches to a competent solar engineer and/or architect. The professional's estimate and the chart's prediction should be within 10—15% of each other.

About the Procedure

In various categories on the chart choose the point value shown if you are close to that condition. For instance, if you have a massive wall but it's not the optimum thickness recommended in the text, you can take less than the 8 points given to that entry. (The 'super' and 'well' insulated home entries are the two hard choices listed. Take no points for a poorly insulated home.)

Let's work a couple through with explanations.

Step I Totalling Points on the Chart

Our example is a solar greenhouse with 200 square feet of south glazing to be added to 1,200 square foot home in Charleston, South Carolina. The greenhouse is within 15° of true south.

1) **Solar Conditions.** Go to the Degree Day Chart on p. 181. Note the months with *over 200* Degree Days. From the possible sunshine chart on p. 180 *average this period* (Nov.–Mar.) and find 61.6% winter solar conditions. Estimating from the range given, take 17 points 17

2) **Latitude.** Charleston is a little below 33°N. 23 points 23

3) **Collector Tilt.** Calculate the degrees from the horizontal that the *majority* of your south facing glazing will be and take that point value. Our example will have a 60° tilt. 8 points 8

Home Heating Contribution Chart

Category	Max Choices	Point Value
Solar Conditions	1	
20-29%		3
30-39%		4-5
40-49%		6-10
50-59%		11-15
60-69%		16-20
70%+		21
Latitude	1	
+50°		3
49-45°		4-7
44-41°		8-10
40-36°		11-19
35-30°		20-25
Collector Tilt	1	
10-30°		2
31-44°		5
45-60°		8
61-90°		10
Night Insulation for Glazing	1	
Radiant Barrier		5
R 1-3		5
R 4-8		8
R 8-10		10
Relationship with Home	2	
Integral-home surrounds GH		12
Two story high GH		6
Indented on east or west corner		5
GH dropped below frostline		3
Add-on to south wall		3
Protected by wind barrier		2
Wall Between Home and GH	1	
Massive material, optimum thickness—medium to dark color		10

Category	Max Choices	Point Value
Home Construction	2	
Super insulated and winterized exceeds local FHA standards		10
Well insulated-meets local FHA standards		7
Isolated thermal storage under home floor-with fan circulation		6
Interior thermal mass in home-brick floors, fireplace, etc.		3
Air Exchange with Home	2	
Fan assist to adjoining rooms		7
Good communication with home		4
Windows/doors adequate size		3
High/low vents in series		3
Other Factors	3	
GH used as main entry on home		6
GH shut off from home at night		4
Home thermostat off on clear winter days		4
No existing windows at GH site		3
Home heating needs in May and Sept.		3
Air lock on GH		2
For 7000+Degree Day/ and/or 50°+ Latitudes only	2	
Summer heating needs		6
Used with no thermal mass-Dec.-Feb.		5
Constant snow cover in winter		5
TOTAL POINTS		____

4) **Night Insulation.** These points *only* apply if the insulating barrier is well sealed and regularly applied in the coldest months. In Charleston, we really won't need it. 0 points. 0

5) **Relationship with Home.** It will be a simple add-on (3 points) and will be below the negligible frostline (3 points). 6

6) **Wall Between.** No points for our case. We have a standard frame home with no significant thermal mass. Leave the wall a light color for the plants. 0

7) **Home Construction.** It's a well (but not super) insulated home. (7 points.) We aren't going to build in a rockbed and don't have any other interior thermal mass in communication with the greenhouse. 7

8) **Air Exchange.** A fan in an adjoining window is planned (7 points) and the greenhouse has good air communication with most of the home (4 points). 11

9) **Other Factors.** Some of these entries involve home management as much as greenhouse design, but they are very effective in boosting performance. For example, using the greenhouse as an airlock is worth 6 points. If you can get your family to use it half the time in winter, take 3 points. For this example, let's say that this isn't possible. We will take two of the available three choices: 'greenhouse shut-off from house at night' (4 points) and 'thermostat off on clear winter days' (4 points) 8

10) **7000 DD+/50^{0} Latitude +.** Two choices are available to cold climate dwellers but our Charleston example isn't even close to this category. 0

TOTAL POINTS 80

Step II Degree Day Factor

From the list below estimate the factor that matches your degree days. (Chart p. 181)

Degree Days	*Degree Day Factor*
2000	1.20
3000	1.17
4000	1.14
5000	1.11
6000	1.08
7000	1.05

Charleston has 2033 degree days which would be a 1.20 degree day factor. A location with 2500 DD would be about 1.185, 2850 is 1.175, etc.

Multiply the points by the Degree Day Factor.

Points X Degree Day Factor = New Points

80 X 1.2 = 96

Step III Orientation Factor

From the list below multiply the new points by the closest orientation factor. Orientation here is the number of degrees from true south your greenhouse/collector glazing faces.

Degrees from South	*Orientation Factor*
0–15°	1
15–30°	.94
30–40°	.87
40–50°	.80
50–60°	.73
60–70°	.65
70–80°	.50
80–90°(Due east or west)	.36
90°+	Find another place to build

Our example is within 15° of true south.

New points	X	Orientation Factor	=	Total point value
96	X	1	=	96

Step IV The Heating Value

From the Key below, choose the range in which your total point value falls.

Key

Point Value	*Range*
120–135	4+
105–120	3+
90–105	2+
75–90	1+
50–75	0–1

96 points puts the example in the 2+ range. By interpolation we will find the exact place in this range.

The 2+ range contains 15 points. (105−90=15) The low point of the range is 90 and our place, 96, is 6 points above that. (96−90=6) Divide 6 by the points in the range, 15. (6÷15=.40) Add the .40 to the range, 2, to get a heating factor; 2.40.

The heating factor is then multiplied by the square footage of south glazing in the greenhouse. Our example has 200 square feet of south glazing.

Heating factor	X	South Glazing	=	Heating Value (in square feet of home heated)
2.4	X	200	=	480

Step V The Solar Fraction

To find out how the greenhouse will effect your home heating needs, divide the heating value by the total square footage of the home.

For the example.

Heating Value	÷	Square Footage of Home	=	Solar Fraction
480	÷	1,200	=	.40 or 40%

In this case, the solar greenhouse in Charleston will take care of about 40% of the total heating requirements for the 1,200 square foot home.

Now, let's move the same basic greenhouse and house to a colder climate, Boston, Mass., and see how it does.

Step I Totalling Points on the Chart

1) **Solar Conditions.** Averaging October through May we get 53.6% 12
2) **Latitude.** 42^oN. 9
3) **Collector Tilt.** Same . . . 60° 8
4) **Night Insulation.** No 0
5) **Relationship with home.** Same. 6
6) **Wall Between.** Same. 0
7) **Home Construction.** Same 7
8) **Air Exchange.** Same 11
9) **Other Factors.** We will shut it off from the home at night (4) but we're not certain that we'll turn off the home thermostat on clear winter days. So, no points for that entry. We do have slightly over 200 degree days in May. (3 points) 7
10) **7000+Degree Days/50°+ Latitude.** No 0

TOTAL POINTS 60

Step II Degree Day Factor

From the list on p. 181-2 Boston, with 5634 degree days (from chart p. 181) has a degree day factor of about 1.09.

Points	X	Degree Day Factor	=	New Points
60	X	1.09	=	65.4

Step III Orientation Factor

Our greenhouse is within 15° of south and the list on p. 170 gives that orientation factor as 1.

New Points	X	Orientation Factor	=	Total Point Value
65.4	X	1	=	65.4

Step IV Heating Value

The Key on p. 170 shows 65.4 in the lowest range, 0–1. Interpolation gives us a heating factor of .616 in that range.

Heating Factor	X	Square Feet of Greenhouse Glazing	=	Heating Value
.616	X	200	=	123.2

Step V Solar Fraction

For the solar fraction we simply divide the Heating Value by the total square feet of the home.

Heating Value	÷	Square Footage of Home	=	Solar Fraction
123.2	÷	1200	=	.102 or 10%

A 10% contribution will help, but we can greatly improve it. Suppose and R-5 insulating barrier was added (8 points) and we found a way to use the greenhouse as the primary entrance for the home. (6 points) This would bring our Step I up 14 points and would result in a 23% solar fraction, or nearly ¼ of our home heating . Not bad for a small greenhouse in Boston!

APPENDIX B

SUN MOVEMENT CHARTS

The charts can be used to determine:

1) The hours of direct sunlight your greenhouse will receive at any time of the year.
2) How obstructions will shade the unit.
3) The altitude of the sun at any moment.
4) A visualization of the sun's rays striking a flat or tilted plane.

You will need:

1) A small piece of graph paper.
2) A cheap protractor.
3) A straight edge.

Some definitions are in order:

1) **Sun Path:** The apparent (from our viewpoint) movement of the sun through the heavens. On our charts, the sweeping east-to-west lines are the sun's path on the 21st or 22nd of each month.

2) **Altitude:** The height in degrees of the sun *from* a true horizon. Altitudes are shown on the *concentric circles* at 9° intervals in the upper right of the charts.

3) **Azimuth:** The distance in degrees east or west of true south shown on the *radii* of the charts.

4) **Hour-of-day:** The nearly vertical lines represent the solar (not time zone or daylight savings) times of day. They are noted across the top of the sun path line.

1) Find the chart nearest your latitude.

2) On a piece of graph paper draw a scale model of the floor plan of your greenhouse. A size of *about* one half by one inch fits easily on the chart. Cut it out. You may want to cut out scale drawings of your home and any obstructions. To be accurate, all models must be measured and positioned to the same scale.

3) On the greenhouse model mark the junction of all solid and clear walls.

4) Place the model on the chart. By using the azimuth angles, position the model in its actual orientation. In our example, the site is facing true south so the model is parallel to the 90° east–90° west azimuth line. The greenhouse center should be in the exact center of the chart. If you have included a house, trees or other obstructions orient them to scale and position them on the chart.

5) The first objective is to examine the duration of sunlight the greenhouse will receive at various times of the year. Choose a month and follow its sun path, noting the number of hours the model is receiving direct sunlight. Solid walls and obstructions will shade the incoming light at certain times of day (see Figure 210). Observing the hours of shading will allow you to design clear and solid walls to best suit your particular location.

6) Next, to determine any solar altitude, find the point at which a chosen hour-of-day line intersects the sun path line. Now find the nearest *concentric* circle and follow it around to the degree marking in the upper right. That's the sun's altitude. When you're in between circles, estimate.

7) To better visualize the angle of incoming light, place the flat edge of a protractor across the time-of-day/sun path intersection you are studying with the midle of the protractor over the center of the chart. Take a straight edge and connect the center of the protractor to the determined solar altitude angle on the edge. That's where the sun is at in that moment of time (see Fig. 212).

8) A side view of the greenhouse cut out of paper is helpful in determining light patterns through clear roof areas and sides (Fig. 202). By repositioning the solid/clear areas in the model, you should be able to get maximum winter sunlight for your location and also obtain some summer shading.

9) This same procedure can be used for any solar application.

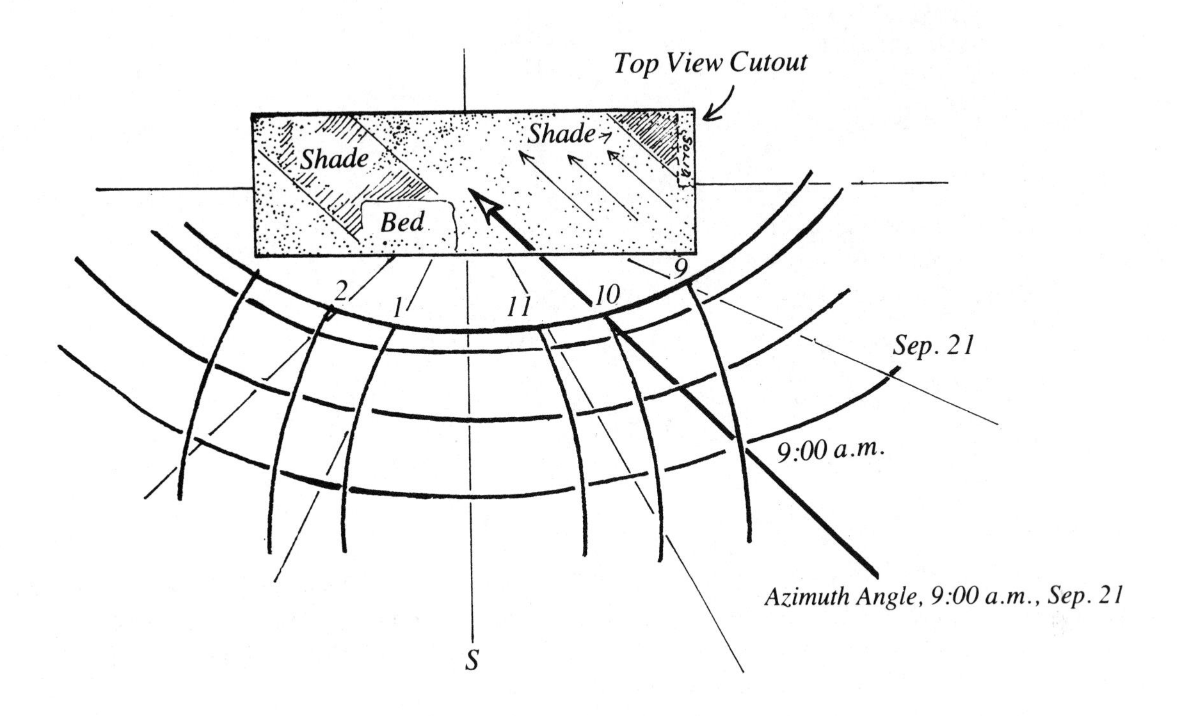

Figure 201

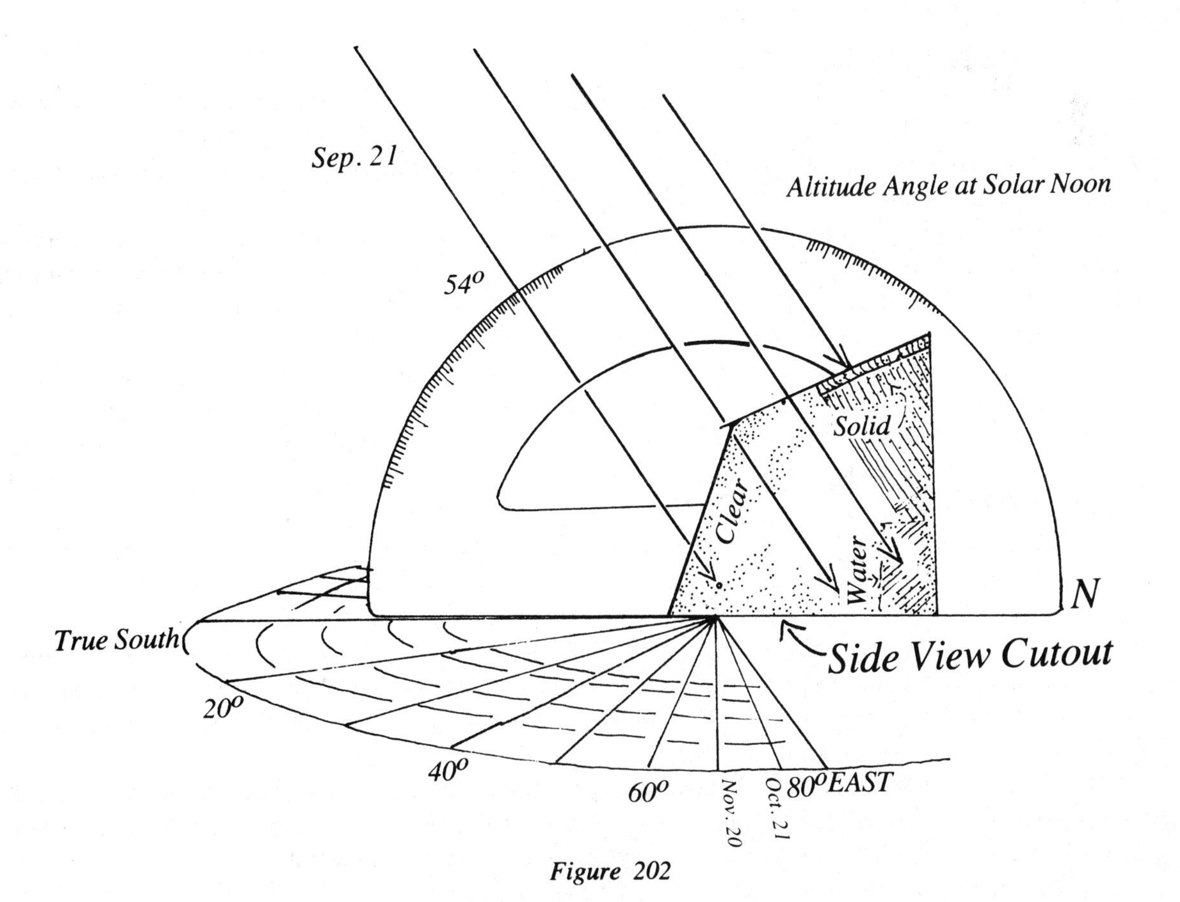

Figure 202

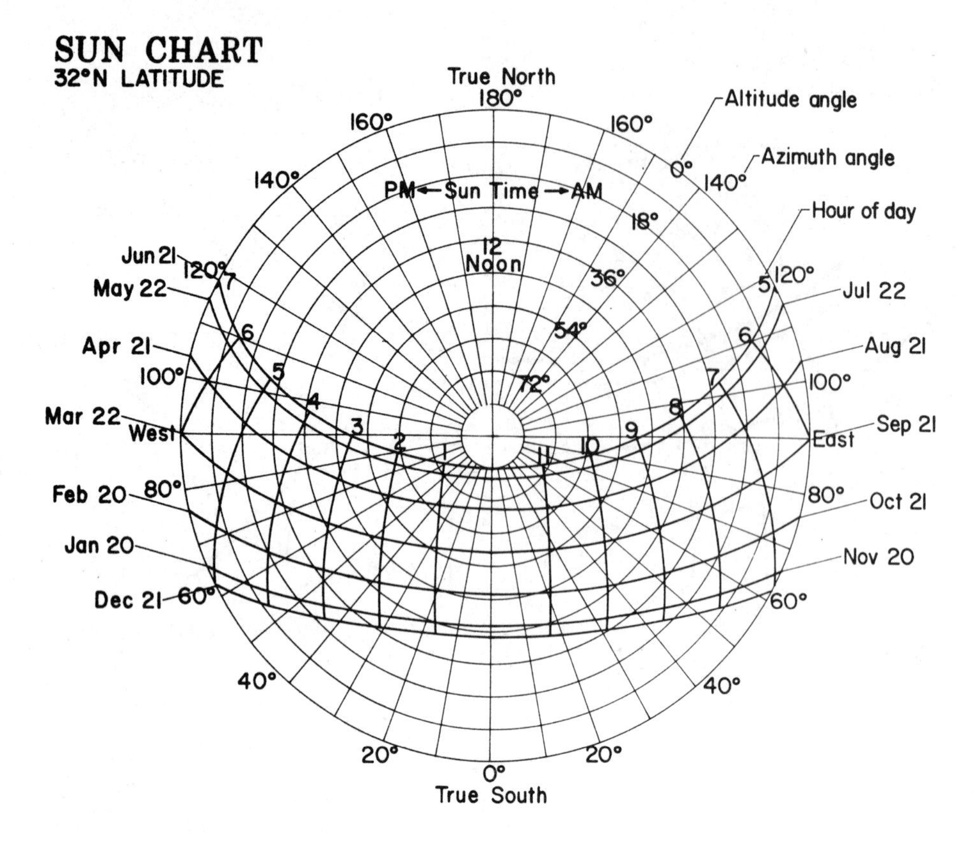

Figure 203

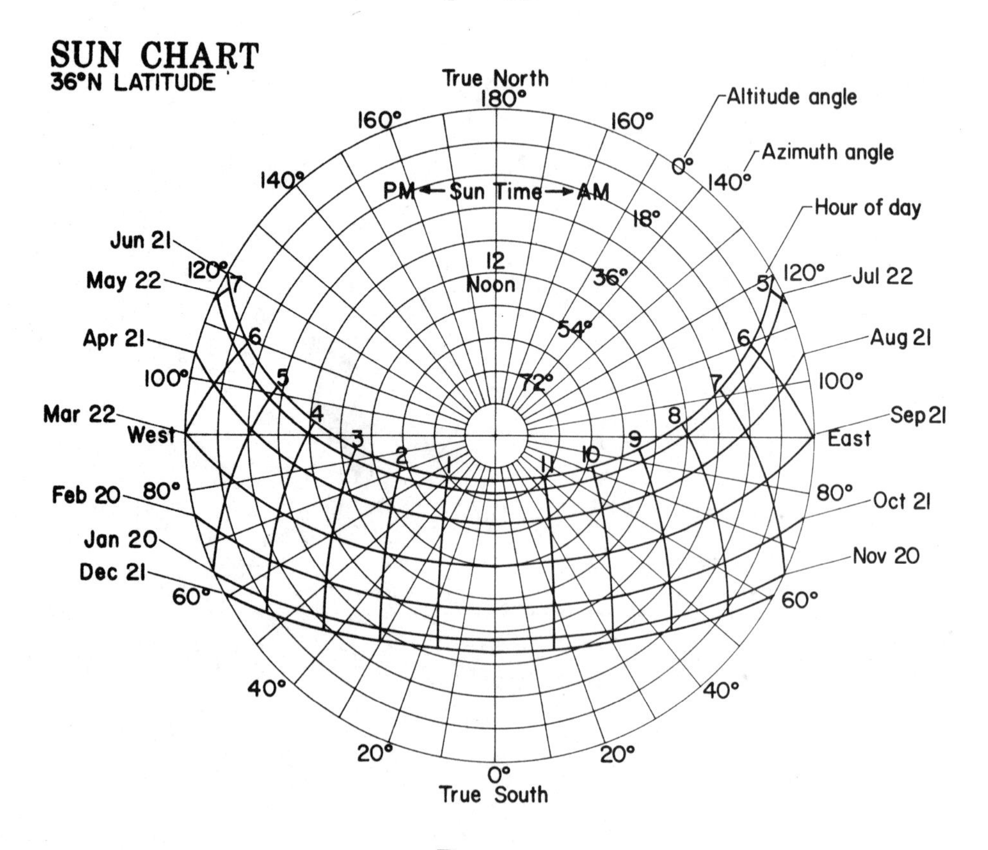

Figure 204

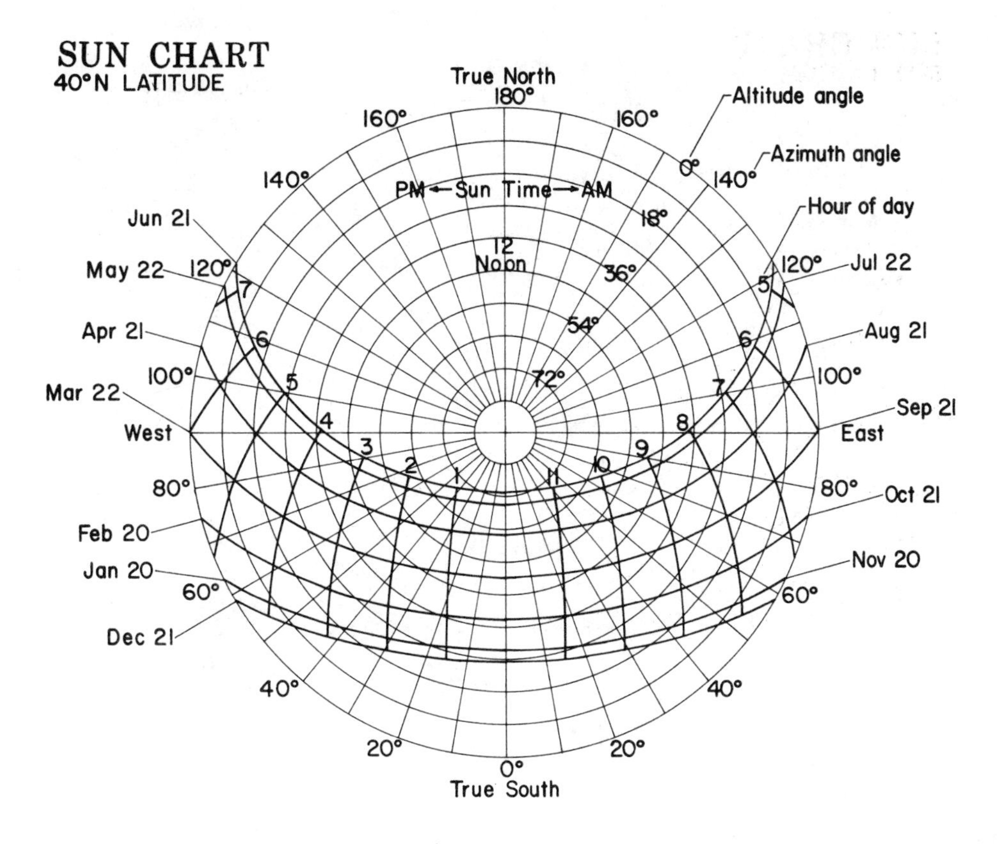

Figure 205

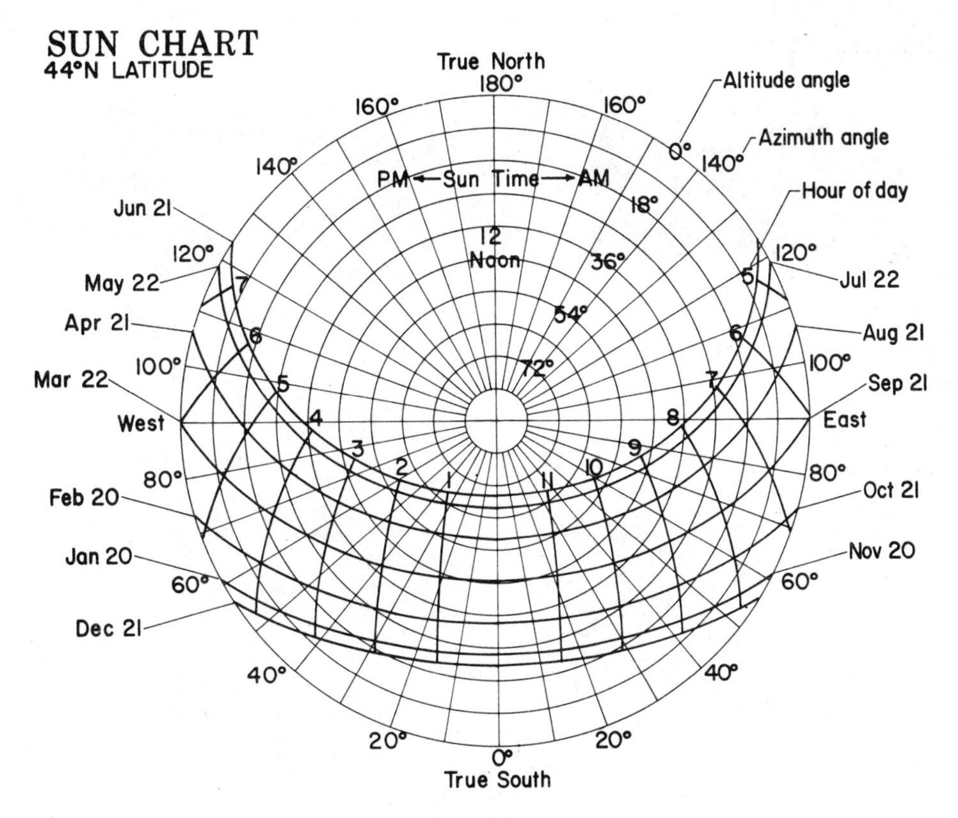

Figure 206

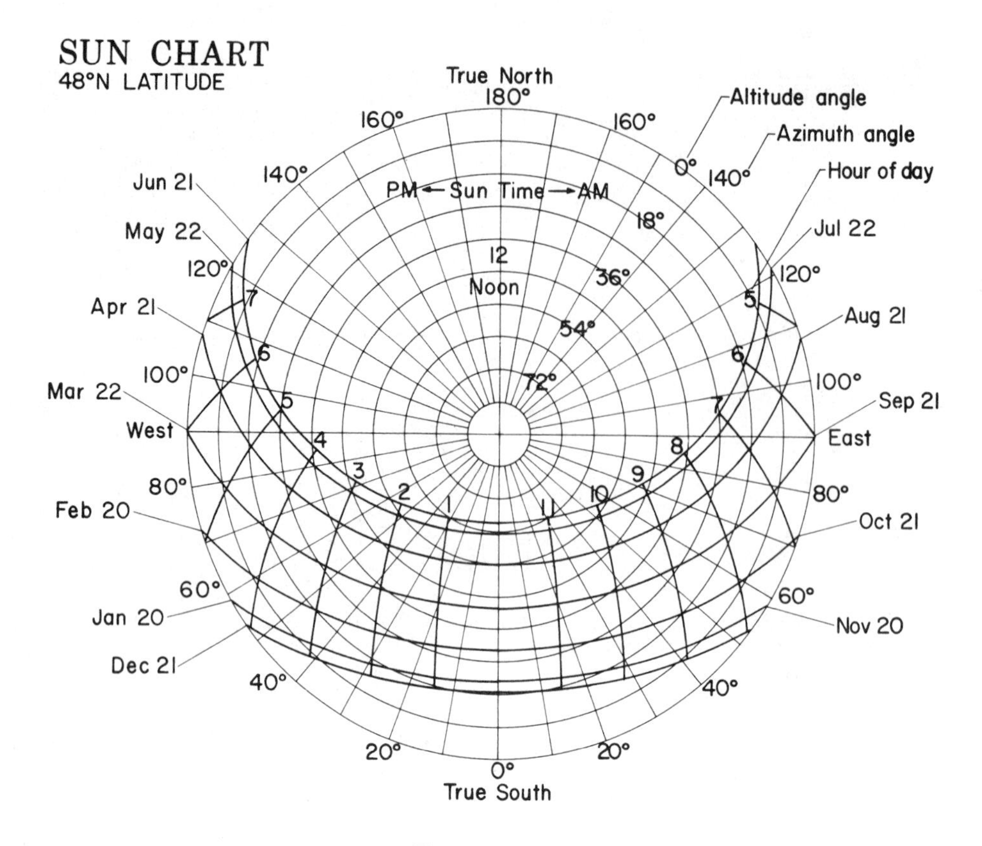

Figure 207

The photographs that follow illustrate the procedure for finding azimuth and altitude angles described above.

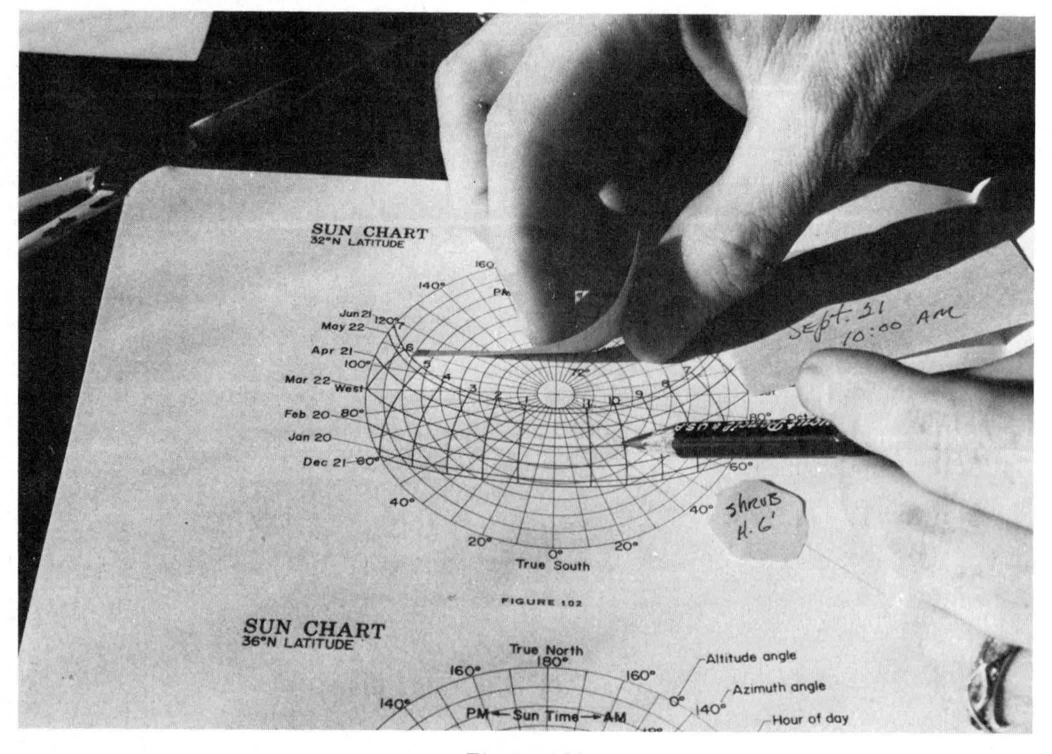

Figure 208

Figure 208. Graph paper cutout of greenhouse plan is oriented on chart. Point at which hour of day (10:00 a.m.) and day of year (Sept. 21) intersect is located.

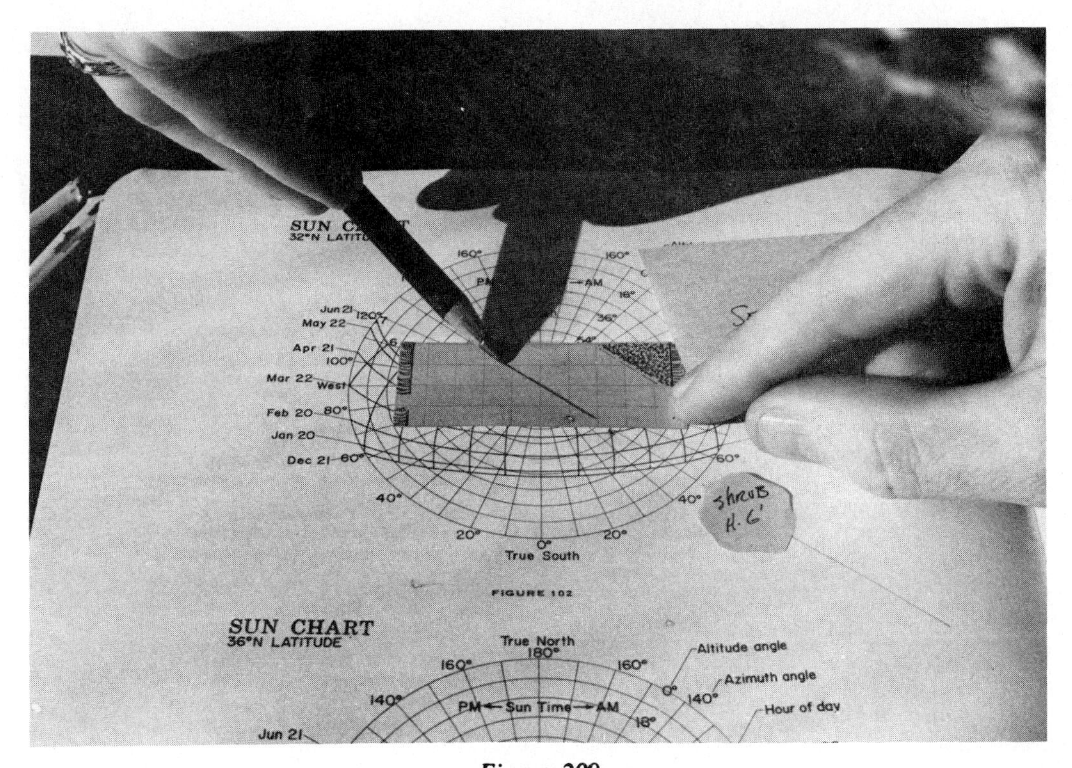

Figure 209

Figure 209. Azimuth line (radii) closest to the above intersection is noted and extended onto the cut out.

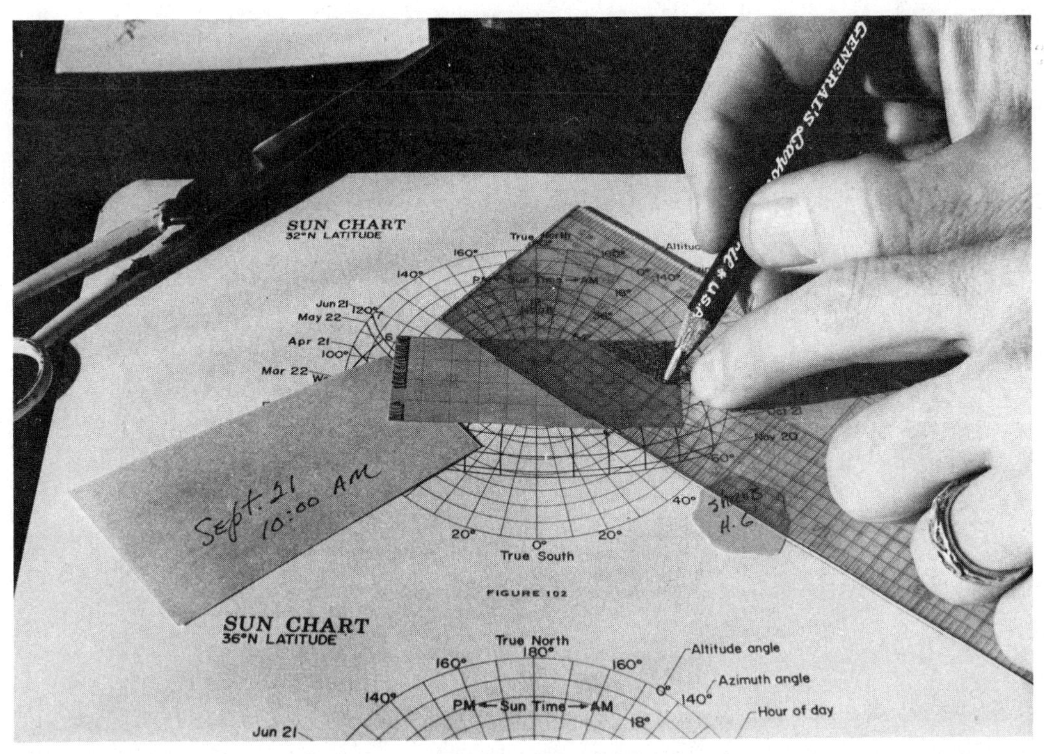

Figure 210

Figure 210. Shadow cast by partial east wall will run parallel to the azimuth line at this time of day.

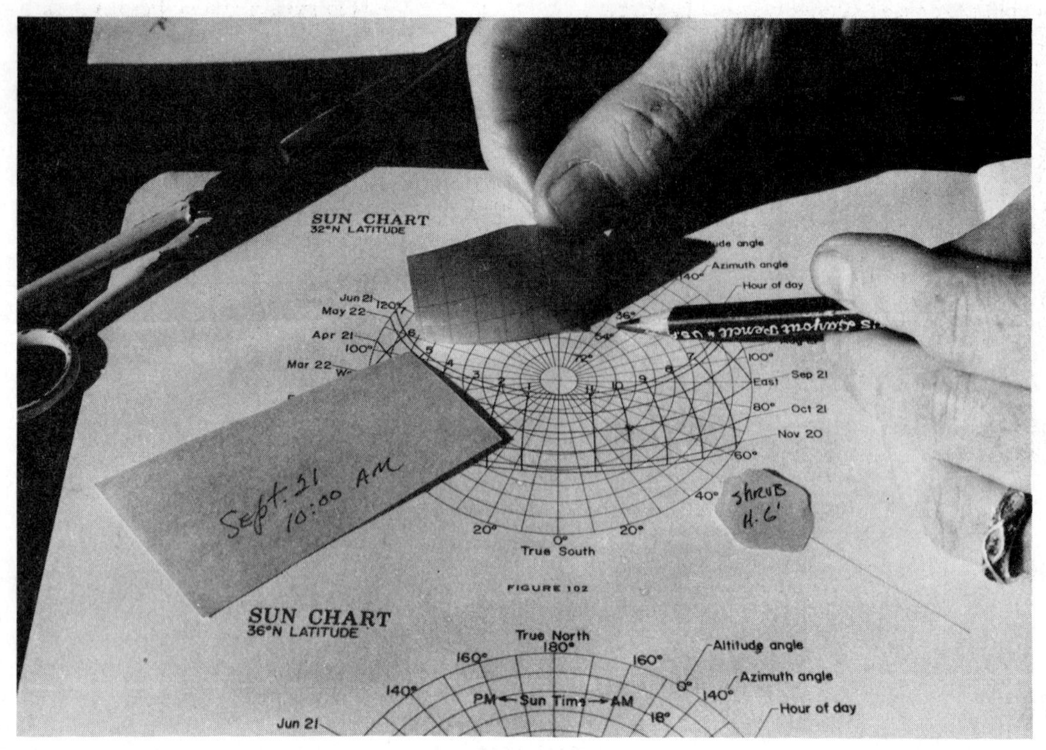

Figure 211

Figure 211. To determine the length of the shadow and shade from solid roof section, locate altitude-angle circle closest to intersection of hour and day.

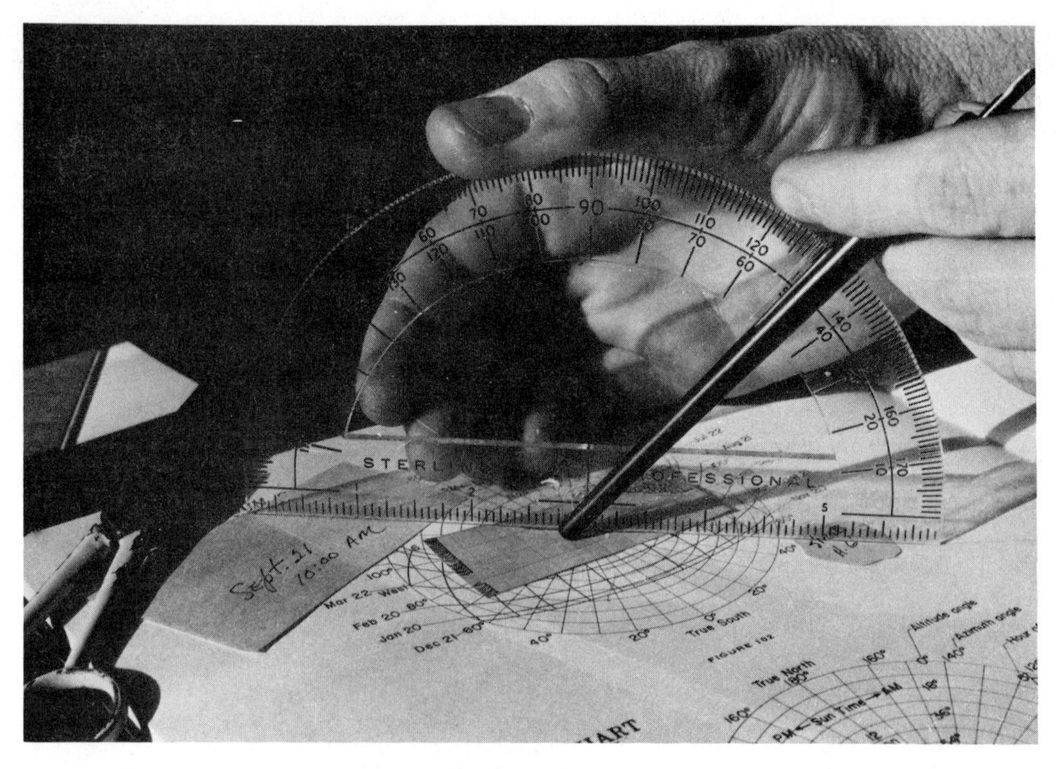

Figure 212

Figure 212. Noting the altitude angle (45°) on the chart, center your protractor over the cutout (along the azimuth line) and find the altitude angle on it.

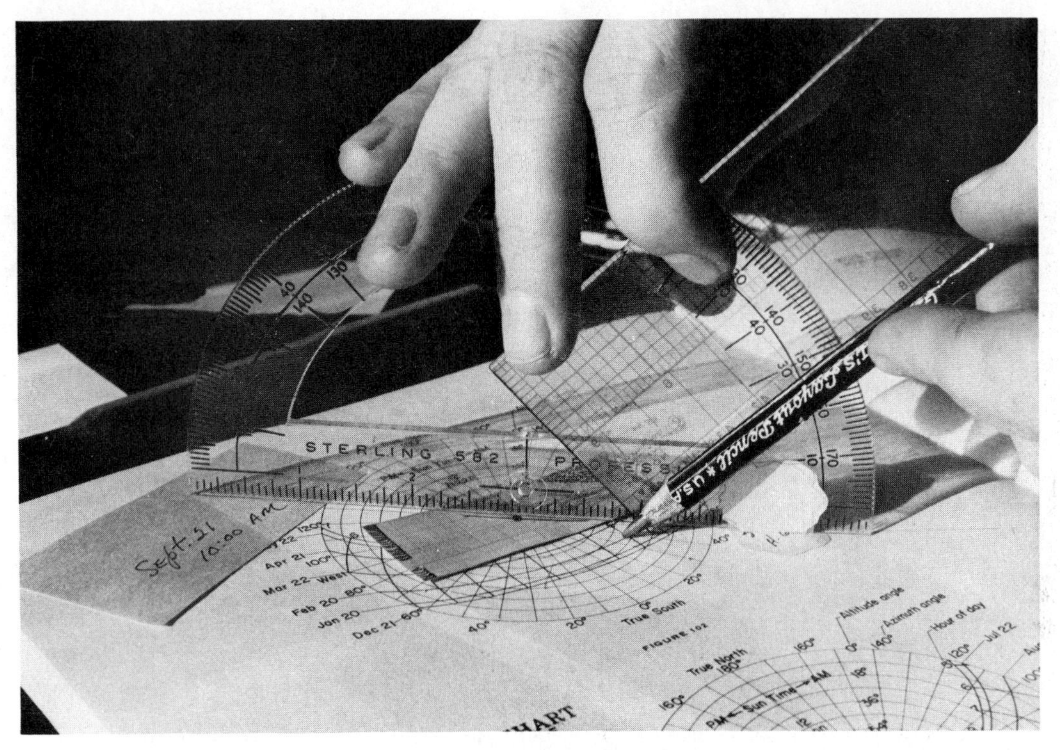

Figure 213

Figure 213. Simulated light ray (the pencil) running parallel to this altitude angle indicates the length of shadow cast by scale cutout of obstruction.

Note that at this hour on the fall/spring equinoxes the shadow falls short of the greenhouse and would therefore not shade the interior. On December 21, however, the altitude angle at 10:00 a.m. would be about 25°. The shrub would then present a shading problem, at the time of year when maximum light and heat are essential.

APPENDIX C

AVERAGE PERCENTAGE OF POSSIBLE SUNSHINE FOR SELECTED LOCATIONS

LOCATION	JAN.	FEB.	MAR.	APR.	MAY	JUN.	JUL.	AUG.	SEP.	OCT.	NOV.	DEC.	ANN.
ALA. BIRMINGHAM	43	49	56	63	66	67	62	65	66	67	58	44	59
MONTGOMERY	51	53	61	69	73	72	66	69	69	71	64	48	64
ALASKA, ANCHORAGE	39	46	56	58	50	51	45	39	35	32	33	29	45
FAIRBANKS	34	50	61	68	55	53	45	35	31	28	38	29	44
JUNEAU	30	32	39	37	34	35	28	30	25	18	21	18	30
NOME	44	46	48	53	51	48	32	26	34	35	36	30	41
ARIZ. PHOENIX	76	79	83	88	93	94	84	84	89	88	84	77	85
YUMA	83	87	91	94	97	98	92	91	93	93	90	83	91
ARK. LITTLE ROCK	44	53	57	62	67	72	71	73	71	74	58	47	62
CALIF. EUREKA	40	44	50	53	54	56	51	46	52	48	42	39	49
FRESNO	46	63	72	83	89	94	97	97	93	87	73	47	78
LOS ANGELES	70	69	70	67	68	69	80	81	80	76	79	72	73
RED BLUFF	50	60	65	75	79	86	95	94	89	77	64	50	75
SACRAMENTO	44	57	67	76	82	90	96	95	92	82	65	44	77
SAN DIEGO	68	67	68	66	60	60	67	70	70	70	76	71	68
SAN FRANCISCO	53	57	63	69	70	75	68	63	70	70	62	54	66
COLO. DENVER	67	67	65	63	61	69	68	68	71	71	67	65	67
GRAND JUNCTION	58	62	64	67	71	79	76	72	77	74	67	58	69
CONN. HARTFORD	46	55	56	54	57	60	62	60	57	55	46	46	56
D.C. WASHINGTON	46	53	56	57	61	64	64	62	62	61	54	47	58
FLA. APALACHICOLA	59	62	62	71	77	70	64	63	62	74	66	53	65
JACKSONVILLE	58	59	66	71	71	63	62	63	58	58	61	53	62
KEY WEST	68	75	78	78	76	70	69	71	65	65	69	66	71
MIAMI BEACH	66	72	73	73	68	62	65	67	62	62	65	65	67
TAMPA	63	67	71	74	75	66	61	64	64	67	67	61	68
GA. ATLANTA	48	53	57	65	68	68	62	63	65	67	60	47	60
HAWAII. HILO	48	42	41	34	31	41	44	38	42	41	34	36	39
HONOLULU	62	64	60	62	64	66	67	70	70	68	63	60	65
LIHUE	48	48	48	46	51	60	58	59	67	58	51	49	54
IDAHO. BOISE	40	48	59	67	68	75	89	86	81	66	46	37	66
POCATELLO	37	47	58	64	66	72	82	81	78	66	48	36	64
ILL. CAIRO	46	53	59	65	71	77	82	79	75	73	56	46	65
CHICAGO	44	49	53	56	63	69	73	70	65	61	47	41	59
SPRINGFIELD	47	51	54	58	64	69	76	72	73	64	53	45	60
IND. EVANSVILLE	42	49	55	61	67	73	78	76	73	67	52	42	64
FT. WAYNE	38	44	51	55	62	69	74	69	64	58	41	38	57
INDIANAPOLIS	41	47	49	55	62	68	74	70	68	64	48	39	59
IOWA. DES MOINES	56	56	56	59	62	66	75	70	64	64	53	48	62
DUBUQUE	48	52	52	58	60	63	73	67	61	55	44	40	57
SIOUX CITY	55	58	58	59	63	67	75	72	67	65	53	50	63
KANS. CONCORDIA	60	60	62	63	65	73	79	76	72	70	64	58	67
DODGE CITY	67	66	68	68	68	74	78	78	76	75	70	67	71
WICHITA	61	63	64	64	66	73	80	77	73	69	67	59	69
KY. LOUISVILLE	41	47	52	57	64	68	72	69	68	64	51	39	59
LA. NEW ORLEANS	49	50	57	63	66	64	58	60	64	70	60	46	59
SHREVEPORT	48	54	58	60	69	78	79	80	79	77	65	60	69
MAINE, EASTPORT	45	51	52	52	51	53	55	57	54	50	37	40	50
MASS. BOSTON	47	56	57	56	59	62	64	63	61	58	48	48	57
MICH. ALPENA	29	43	52	56	59	64	70	64	52	44	24	22	51
DETROIT	34	42	48	52	58	65	69	66	61	54	35	29	53
GRAND RAPIDS	26	37	48	54	60	66	72	67	58	50	31	22	49
MARQUETTE	31	40	47	52	53	56	63	57	47	38	24	24	47
S. STE. MARIE	28	44	50	54	54	59	63	58	45	36	21	22	47
MINN. DULUTH	47	55	60	58	58	60	68	63	53	47	36	40	55
MINNEAPOLIS	49	54	55	57	60	64	72	69	60	54	40	40	56
MISS. VICKSBURG	46	50	57	64	69	73	69	72	74	71	60	45	64
MO. KANSAS CITY	55	57	59	60	64	70	76	73	70	67	59	52	65
ST. LOUIS	48	49	56	59	64	68	72	68	67	65	54	44	61
SPRINGFIELD	48	54	57	60	53	69	77	72	71	65	58	48	63
MONT. HAVRE	49	58	61	63	63	65	78	75	64	57	48	46	62
HELENA	46	55	58	60	59	63	77	74	63	57	48	43	60
KALISPELL	28	40	49	57	58	60	77	73	61	50	28	20	53
NEBR. LINCOLN	57	59	60	60	63	69	76	71	67	66	59	55	64
NORTH PLATTE	63	63	64	62	64	72	78	74	72	70	62	58	68
NEV. ELY	61	64	68	65	67	79	79	81	81	73	67	62	72
LAS VEGAS	74	77	78	81	85	91	84	86	92	84	83	75	82
RENO	59	64	69	75	77	82	90	89	86	76	68	56	76
WINNEMUCCA	52	60	64	70	76	83	90	90	86	75	62	53	74
N.H. CONCORD	48	53	55	53	51	56	57	58	55	50	43	43	52
N.J. ATLANTIC CITY	51	57	58	59	62	65	67	66	65	54	58	52	60
N. MEX. ALBUQUERQUE	70	72	72	76	79	84	76	75	81	80	79	70	76
ROSWELL	69	72	75	77	76	80	76	75	74	74	74	69	74
N.Y. ALBANY	43	51	53	53	57	62	63	61	58	54	39	38	53
BINGHAMTON	31	39	41	44	50	56	54	51	47	43	29	26	44
BUFFALO	32	41	49	51	59	67	70	67	60	51	31	28	53
CANTON	37	47	50	48	54	61	63	61	54	45	30	31	49
NEW YORK	49	56	57	59	62	65	66	64	64	61	53	50	59
SYRACUSE	31	38	45	50	58	64	67	63	56	47	29	26	50
N.C. ASHEVILLE	48	53	56	61	64	63	59	59	62	64	59	48	58
RALEIGH	50	56	59	64	67	65	62	62	63	64	62	52	61
N. DAK. BISMARCK	52	58	56	57	58	61	73	69	62	59	49	48	59
DEVILS LAKE	53	60	59	60	59	62	71	67	59	56	44	45	58
FARGO	47	55	56	58	62	63	73	69	60	57	39	46	59
WILLISTON	51	59	60	63	66	66	78	75	65	60	48	48	63
OHIO, CINCINATTI	41	46	52	56	62	69	72	68	68	60	46	39	57
CLEVELAND	29	36	45	52	61	67	71	68	62	54	32	25	50
COLUMBUS	36	44	49	54	63	68	71	68	66	60	44	35	55
OKLA. OKLAHOMA CITY	57	60	63	64	65	74	78	78	74	68	64	57	68
OREG. BAKER	41	49	56	61	63	67	83	81	74	62	46	37	60
PORTLAND	27	34	41	49	52	55	70	65	55	42	28	23	48
ROSEBURG	24	32	40	51	57	59	79	77	68	42	28	18	51
PA. HARRISBURG	43	52	55	57	61	65	68	63	62	58	47	43	57
PHILADELPHIA	45	56	57	58	61	62	64	61	62	61	53	49	57
PITTSBURG	32	39	45	50	57	62	64	61	62	54	39	30	51
R.I. BLOCK ISLAND	45	54	47	56	58	60	62	62	60	59	50	44	56
S.C. CHARLESTON	58	60	65	72	73	70	66	66	67	68	68	57	66
COLUMBIA	53	57	62	68	69	68	63	65	64	68	64	51	63
S. DAK. HURON	55	62	60	62	65	68	76	72	66	61	52	49	63
RAPID CITY	58	62	63	62	61	66	73	73	69	66	58	54	64
TENN. KNOXVILLE	42	49	53	59	64	66	64	59	64	64	53	41	57
MEMPHIS	44	51	57	64	68	74	73	74	70	69	58	45	64
NASHVILLE	42	47	54	60	65	69	69	68	69	65	55	42	59
TEX. ABILENE	64	68	73	66	73	86	83	85	73	71	72	66	73
AMARILLO	71	71	75	75	75	82	81	81	79	76	76	70	76
AUSTIN	46	50	57	60	62	72	76	79	70	70	57	49	63
BROWNSVILLE	44	49	51	57	65	73	78	78	67	70	54	44	61
DEL RIO	53	55	61	63	60	66	75	80	69	66	58	52	63
EL PASO	74	77	81	85	87	87	78	78	80	82	80	73	80
FT. WORTH	56	57	65	66	67	75	78	78	74	70	63	58	68
GALVESTON	50	50	55	61	69	76	72	71	70	74	62	49	63
SAN ANTONIO	48	51	56	58	60	69	74	75	69	67	55	49	62
UTAH, SALT LAKE CITY	48	53	61	68	73	78	82	82	84	73	56	49	69
VT. BURLINGTON	34	43	48	47	53	59	62	59	51	43	25	24	46
VA. NORFOLK	50	57	60	63	67	66	66	66	63	64	60	51	62
RICHMOND	49	55	59	63	67	66	65	62	63	64	58	50	61
WASH. NORTH HEAD	28	37	42	48	48	48	50	46	48	41	31	27	41
SEATTLE	27	34	42	48	53	48	62	56	53	36	28	24	45
SPOKANE	26	41	53	63	64	68	82	79	68	53	28	22	58
TATOOSH ISLAND	26	36	39	45	47	46	48	44	47	38	26	23	40
WALLA WALLA	24	35	51	63	67	72	86	84	72	59	33	20	60
YAKIMA	34	49	62	70	72	74	86	86	74	61	38	29	65
W. VA. ELKINS	33	37	42	47	55	55	56	53	55	51	41	33	48
PARKERSBURG	30	36	42	49	56	60	63	60	60	53	37	29	48
WIS. GREEN BAY	44	51	55	56	58	64	70	65	58	52	40	40	55
MADISON	44	49	52	53	58	64	70	66	60	56	41	38	56
MILWAUKEE	44	48	53	56	60	65	73	67	62	56	44	39	57
WYO. CHEYENNE	65	66	64	61	59	68	70	68	69	60	65	63	66
LANDER	66	70	71	66	65	74	76	75	72	67	61	62	69
SHERIDAN	56	61	62	61	61	67	76	74	67	60	53	52	64
YELLOWSTONE PARK	39	51	55	57	56	63	73	71	65	57	45	38	56
P.R. SAN JUAN	64	69	71	66	59	62	65	67	61	63	63	65	65

The average monthly percentage of sunshine available in cities throughout the United States.

APPENDIX D

MONTHLY TOTAL HEATING DEGREE DAYS (BASED ON 65 DEGREES F) FOR LOCATIONS AROUND THE NATION.

State & Station	Jul	Aug	Sep	Oct	Nov	Dec	Jan	Feb	Mar	Apr	May	Jun	Annual
Ala. Birmingham	0	0	6	93	363	555	592	462	363	108	9	0	2551
Huntsville	0	0	12	127	426	663	694	557	434	138	19	0	3070
Mobile	0	0	0	22	213	357	415	300	211	42	0	0	1560
Montgomery	0	0	0	68	330	527	543	417	316	90	0	0	2291
Alas Anchorage	245	291	516	930	1284	1572	1631	1316	1293	879	592	315	10864
Annette	242	208	327	567	738	899	949	837	843	648	490	321	7069
Barrow	803	840	1035	1500	1971	2362	2517	2332	2468	1944	1445	957	20174
Barter Is.	735	775	987	1482	1944	2337	2536	2369	2477	1923	1373	924	19862
Bethel	319	394	612	1042	1434	1866	1903	1590	1655	1173	806	402	13196
Cold Bay	474	425	525	772	918	1122	1153	1036	1122	951	791	591	9880
Cordova	366	391	522	781	1017	1221	1299	1086	1113	864	660	444	9764
Fairbanks	171	332	642	1203	1833	2254	2359	1901	1739	1068	555	222	14279
Juneau	301	338	483	725	921	1135	1237	1070	1073	810	601	381	9075
King Salmon	313	322	513	908	1290	1606	1600	1333	1411	966	673	408	11343
Kotzebue	381	446	723	1249	1728	2127	2192	1932	2080	1554	1058	639	16105
McGrath	208	338	633	1184	1791	2232	2294	1817	1758	1122	648	258	14283
Nome	481	496	693	1094	1455	1820	1879	1666	1770	1314	930	573	14171
St. Paul	605	539	612	862	963	1197	1228	1168	1265	1098	936	726	11199
Shemya	577	475	501	784	876	1042	1045	958	1011	885	837	696	9687
Yakutat	338	347	474	716	936	1144	1169	1019	1042	840	632	435	9092
Ariz Flagstaff	46	68	201	558	867	1073	1169	991	911	651	437	180	7152
Phoenix	0	0	0	22	234	415	474	328	217	75	0	0	1765
Prescott	0	0	27	245	579	797	865	711	605	360	158	15	4362
Tucson	0	0	0	25	231	406	471	344	242	75	6	0	1800
Winslow	0	0	6	245	711	1008	1054	770	601	291	96	0	4782
Yuma	0	0	0	0	148	319	363	228	130	29	0	0	1217
Ark Ft. Smith	0	0	12	127	450	704	781	596	456	144	22	0	3292
Little Rock	0	0	9	127	465	716	756	577	438	126	9	0	3219
Texarkana	0	0	0	78	345	561	626	468	350	105	0	0	2533
Ca Bakersfield	0	0	0	37	282	502	546	364	267	105	19	0	2122
Bishop	0	0	42	248	576	797	874	666	539	306	143	36	4227
Blue Canyon	34	50	120	347	579	766	865	781	791	582	397	195	5507
Burbank	0	0	6	43	177	301	366	277	239	138	81	18	1646
Eureka	270	257	258	329	414	499	546	470	505	438	372	285	4643
Fresno	0	0	0	78	339	558	586	406	319	150	56	0	2492
Long Beach	0	0	12	40	156	288	375	297	267	168	90	18	1711
Los Angeles	28	22	42	78	180	291	372	302	288	219	158	81	2061
Mt. Shasta	25	34	123	406	696	902	983	784	738	525	347	159	5722
Oakland	53	50	45	127	309	481	527	400	353	255	180	90	2870
Pt. Arguello	202	186	162	205	291	400	474	392	403	339	298	243	3595
Red Bluff	0	0	0	53	318	555	605	428	341	168	47	0	2515
Sacramento	0	0	12	81	363	577	614	442	360	216	102	6	2773
Sandberg	0	0	30	202	480	691	778	661	620	426	264	57	4209
San Diego	6	0	15	37	123	251	313	249	202	123	84	36	1439
San Francisco	81	78	60	143	306	462	508	395	363	279	214	126	3015
Santa Catalina	16	0	9	50	165	279	353	308	326	249	192	105	2052
Santa Maria	99	93	96	146	270	391	459	370	363	282	233	165	2967
Colo Alamosa	65	99	279	639	1065	1420	1476	1162	1020	696	440	168	8529
Colo Springs	9	25	132	456	825	1032	1128	938	893	582	319	84	6423
Denver	6	9	117	428	819	1035	1132	938	887	558	288	66	6283
Grand Junction	0	0	30	313	786	1113	1209	907	729	387	146	21	5641
Pueblo	0	0	54	326	750	986	1085	871	772	429	174	15	5462
Conn Bridgeport	0	0	66	307	615	986	1079	966	853	510	208	27	5617
Hardfort	0	6	99	372	711	1119	1209	1061	899	495	177	24	6172
New Haven	0	12	87	347	648	1011	1097	991	871	543	245	45	5897
Del Wilmington	0	0	51	270	588	927	980	874	735	387	112	6	4930
Fla Apalachicola	0	0	0	16	153	319	347	260	180	33	0	0	1308
Daytona Beach	0	0	0	0	75	211	248	190	140	15	0	0	879
Ft. Myers	0	0	0	0	24	109	146	101	62	0	0	0	442
Jacksonville	0	0	0	12	144	310	332	246	174	21	0	0	1239
Key West	0	0	0	0	0	28	40	31	9	0	0	0	108
Lakeland	0	0	0	0	57	164	195	146	99	0	0	0	661
Miami Beach	0	0	0	0	0	40	56	36	9	0	0	0	141
Orlando	0	0	0	0	72	198	220	165	105	6	0	0	766
Pensacola	0	0	0	19	195	353	400	277	183	36	0	0	1463
Tallahassee	0	0	0	28	198	360	375	286	202	36	0	0	1485
Tampa	0	0	0	0	60	171	202	143	102	0	0	0	683
W. Palm Beach	0	0	0	0	6	65	87	64	31	0	0	0	253
Ga Athens	0	0	12	115	405	632	642	529	431	141	22	0	2929
Atlanta	0	0	18	127	414	626	639	529	437	168	25	0	2983
Augusta	0	0	0	78	333	552	549	445	350	90	0	0	2397
Columbus	0	0	0	87	333	543	552	434	338	96	0	0	2383

State & Station	Jul	Aug	Sep	Oct	Nov	Dec	Jan	Feb	Mar	Apr	May	Jun	Annual
Macon	0	0	0	71	297	502	505	403	295	63	0	0	2136
Rome	0	0	24	161	474	701	710	577	468	177	34	0	3326
Savannah	0	0	0	47	246	437	437	353	254	45	0	0	1819
Thomasville	0	0	0	25	198	366	394	305	208	33	0	0	1529
Idaho Boise	0	0	132	415	792	1017	1113	854	722	438	245	81	5809
Id Falls 46W	16	34	270	623	1056	1370	1538	1249	1085	651	391	192	8475
Id Falls 42NW	16	40	282	648	1107	1432	1600	1291	1107	657	388	192	8760
Lewiston	0	0	123	403	756	933	1063	815	694	426	239	90	5542
Pocatello	0	0	172	493	900	1166	1324	1058	905	555	319	141	7033
Ill Cairo	0	0	36	164	513	791	856	680	539	195	47	0	3821
Chicago	0	0	81	326	753	1113	1209	1044	890	480	211	48	6155
Moline	0	9	99	335	774	1181	1314	1100	918	450	189	39	6408
Peoria	0	6	87	326	759	1113	1218	1025	849	426	183	33	6025
Rockford	6	9	114	400	837	1221	1333	1137	961	516	236	60	6830
Springfield	0	0	72	291	696	1023	1135	935	769	354	136	18	5429
Ind Evansville	0	0	66	220	606	896	955	767	620	238	68	0	4435
Ft. Wayne	0	9	105	378	783	1135	1178	1028	890	471	189	39	6205
Indianapolis	0	0	90	316	723	1051	1113	949	809	432	177	39	5699
So. Bend	0	6	111	372	777	1125	1221	1070	933	525	239	60	6439
Iowa Burlington	0	0	93	322	768	1135	1259	1042	859	426	177	33	6114
Des Moines	0	9	99	363	837	1231	1398	1165	967	489	211	39	6808
Dubuque	12	31	156	450	906	1287	1420	1204	1026	546	260	78	7376
Sioux City	0	9	108	369	867	1240	1435	1198	989	483	214	39	6951
Waterloo	12	19	138	428	909	1296	1460	1221	1023	531	229	54	7320
Kan Concordia	0	0	57	276	705	1023	1163	935	781	372	149	18	5479
Dodge City	0	0	33	251	666	939	1051	840	719	354	124	9	4986
Goodland	0	6	81	381	810	1073	1166	955	884	507	236	42	6141
Topeka	0	0	57	270	672	980	1122	893	722	330	124	12	5182
Wichita	0	0	33	229	618	905	1023	804	645	270	87	6	4620
Ky Covington	0	0	75	291	669	983	1035	893	756	390	149	24	5265
Lexington	0	0	54	239	609	902	946	818	685	325	105	0	4683
Louisville	0	0	54	248	609	890	930	818	682	315	105	9	4660
La Alexandria	0	0	0	56	273	431	471	361	260	69	0	0	1921
Baton Rouge	0	0	0	31	216	369	409	294	208	33	0	0	1560
Burrwood	0	0	0	0	96	214	298	218	171	27	0	0	1024
Lake Charles	0	0	0	19	210	341	381	274	195	39	0	0	1459
New Orleans	0	0	0	19	192	322	363	258	192	39	0	0	1385
Shreveport	0	0	0	47	297	477	552	426	304	81	0	0	2184
Me Caribou	78	115	336	682	1044	1535	1690	1470	1308	858	468	183	9767
Portland	12	53	195	508	807	1215	1339	1182	1042	675	372	111	7511
Md Baltimore	0	0	48	264	585	905	936	820	679	327	90	0	4654
Frederick	0	0	66	307	624	955	995	876	741	384	127	12	5087
Ma BlueHil Obsy	0	22	108	381	690	1085	1178	1053	936	579	267	69	6368
Boston	0	9	60	316	603	983	1088	972	846	513	208	36	5634
Nantucket	12	22	93	332	573	896	992	941	896	621	384	129	5891
Pittsfield	25	59	219	524	831	1231	1339	1196	1063	660	326	105	7578
Worcester	6	34	147	450	774	1172	1271	1123	998	612	304	78	6969
Mich Alpena	68	105	273	580	912	1268	1404	1299	1218	777	446	156	8506
Detroit (City)	0	0	87	360	738	1088	1181	1058	936	522	220	42	6232
Escanaba	59	87	243	539	924	1293	1445	1296	1203	777	456	159	8481
Flint	16	40	159	465	843	1212	1330	1198	1066	639	319	90	7377
Grand Rapids	9	28	135	434	804	1147	1259	1134	1011	579	279	75	6894
Lansing	6	22	138	435	813	1163	1262	1142	1011	579	273	69	6909
Marquette	59	81	240	527	936	1268	1411	1268	1187	771	468	177	8393
Muskegon	12	28	120	400	762	1088	1209	1100	995	594	310	78	6696
Sault Ste Marie	96	105	279	580	951	1367	1525	1380	1277	810	477	201	9048
Minn Duluth	71	109	330	632	1131	1581	1745	1518	1355	840	490	198	10000
Intnl Falls	71	112	363	701	1236	1724	1919	1621	1414	828	443	174	10606
Minneapolis	22	31	189	505	1014	1454	1631	1380	1166	621	288	81	8382
Rochester	25	34	186	474	1005	1438	1593	1366	1150	630	301	93	8295
St. Cloud	28	47	225	549	1065	1500	1702	1445	1221	666	326	105	8879
Miss Jackson	0	0	0	65	315	502	546	414	310	87	0	0	2239
Meridian	0	0	0	81	339	518	543	417	310	81	0	0	2289
Vicksburg	0	0	0	53	279	462	512	384	282	69	0	0	2041
Mo Columbia	0	0	54	251	651	967	1076	874	716	324	121	12	5046
Kansas	0	0	39	220	612	905	1032	818	682	294	109	0	4711
St. Joseph	0	6	60	285	708	1039	1172	949	769	348	133	15	5484
St. Louis	0	0	60	251	627	936	1026	848	704	312	121	15	4900
Springfield	0	0	45	223	600	877	973	781	660	291	105	6	4561
Mont Billings	6	15	186	487	897	1135	1296	1100	970	570	285	102	7049
Glasgow	31	47	270	608	1104	1466	1711	1439	1187	648	335	150	8996
Great Falls	28	53	258	543	921	1169	1349	1154	1063	642	384	186	7750

State & Station	Jul	Aug	Sep	Oct	Nov	Dec	Jan	Feb	Mar	Apr	May	Jun	Annual
Havre	28	53	306	595	1065	1367	1584	1364	1181	657	338	162	8700
Helena	31	59	294	601	1002	1265	1438	1170	1042	651	381	195	8129
Kalispell	50	99	321	654	1020	1240	1401	1134	1029	639	397	207	8191
Miles City	6	6	174	502	972	1296	1504	1252	1057	579	276	99	7723
Missoula	34	74	303	651	1035	1287	1420	1120	970	621	391	219	8125
Neb Grand Is.	0	6	108	381	834	1172	1314	1089	908	462	211	45	6530
Lincoln	0	6	75	301	726	1066	1237	1016	834	402	171	30	5864
Norfolk	9	0	111	397	873	1234	1414	1179	983	498	233	48	6979
North Platte	0	6	123	440	885	1166	1271	1039	930	519	248	57	6684
Omaha	0	12	105	357	828	1175	1355	1126	939	465	208	42	6612
Scottsbluff	0	0	138	459	876	1128	1231	1008	921	552	285	75	6673
Valentine	9	12	165	493	942	1237	1395	1176	1045	579	288	84	7424
Nev Elko	9	34	225	561	924	1197	1314	1036	911	621	409	192	7433
Ely	28	43	234	592	939	1184	1308	1075	977	672	456	225	7733
Las Vegas	0	0	0	78	387	617	688	487	335	111	6	0	2709
Reno	43	87	204	490	801	1026	1073	823	729	510	357	189	6332
Winnemucca	0	34	210	536	876	1091	1172	916	837	573	363	153	6761
N.H. Concord	6	50	177	505	822	1240	1358	1184	1032	636	298	75	7383
Mt.Wash Obsy	493	536	720	1057	1341	1742	1820	1663	1652	1260	630	603	13817
N.J. Atlantic City	0	0	39	251	549	880	936	848	741	420	133	15	4812
Newark	0	0	30	248	573	921	983	876	729	381	118	0	4859
Trenton	0	0	57	264	576	924	989	885	753	399	121	12	4980
N.M. Albuq.	0	0	12	229	642	868	930	703	595	288	81	0	4348
Clayton	0	6	66	310	699	899	986	812	747	429	183	21	5158
Raton	9	28	126	431	825	1048	1116	904	834	543	301	63	6228
Roswell	0	0	18	202	573	806	840	641	481	201	31	0	3793
Silver City	0	0	6	183	525	729	791	605	518	261	87	0	3705
N.Y. Albany	0	19	138	440	777	1194	1311	1156	992	564	239	45	6875
Binghamton	22	65	201	471	810	1184	1277	1154	1045	645	313	99	7286
Buffalo	19	37	141	440	777	1156	1256	1145	1039	645	329	78	7062
Central Park	0	0	30	233	540	902	986	885	760	408	118	9	4871
JFK Intl	0	0	36	248	564	933	1029	935	815	480	167	12	5219
LaGuardia	0	0	27	223	528	887	973	879	750	414	124	6	4811
Rochester	9	31	126	415	747	1125	1234	1123	1014	597	279	48	6748
Schenectady	0	22	123	422	756	1159	1283	1131	970	543	211	30	6650
Syracuse	6	28	132	415	744	1153	1271	1140	1004	570	248	45	6756
N. C. Asheville	0	0	48	245	555	775	784	683	592	273	87	0	4042
Cape Hatteras	0	0	0	78	273	521	580	518	440	177	25	0	2612
Charlotte	0	0	6	124	438	691	691	582	481	156	22	0	3191
Greensboro	0	0	33	192	513	778	784	672	552	234	47	0	3805
Raleigh	0	0	21	164	450	716	725	616	487	180	34	0	3393
Wilmington	0	0	0	74	291	521	546	462	357	96	0	0	2347
WinstonSalem	0	0	21	171	483	747	753	652	524	207	37	0	3595
N.D. Bismarck	34	28	222	577	1083	1463	1708	1442	1203	645	329	117	8851
Devils Lake	40	53	273	642	1191	1634	1872	1579	1345	753	381	138	9901
Fargo	28	37	219	574	1107	1569	1789	1520	1262	690	332	99	9226
Williston	31	43	261	601	1122	1513	1758	1473	1262	681	357	141	9243
Ohio Akron	0	9	96	381	726	1070	1138	1016	871	489	202	39	6037
Cincinnati	0	0	54	248	612	921	970	837	701	336	118	9	4806
Cleveland	9	25	105	384	738	1088	1159	1047	918	552	260	66	6351
Columbus	0	6	84	347	714	1039	1088	949	809	426	171	27	5660
Dayton	0	6	78	310	696	1045	1097	955	809	429	167	30	5622
Mansfield	9	22	114	397	768	1110	1169	1042	924	543	245	60	6403
Sandusky	0	6	66	313	684	1032	1107	991	868	495	198	36	5796
Toledo	0	16	117	406	792	1138	1200	1056	924	543	242	60	6494
Youngstown	6	19	120	412	771	1104	1169	1047	921	540	248	60	6417
Ok Okla City	0	0	15	164	498	766	868	664	527	189	34	0	3725
Tulsa	0	0	18	158	522	787	893	683	539	213	47	0	3860
Ore Astoria	146	130	210	375	561	679	753	622	636	480	363	231	5186
Burns	12	37	210	515	867	1113	1246	988	856	570	366	177	6957
Eugene	34	34	129	366	585	719	803	627	589	426	279	135	4726
Meacham	84	124	288	580	918	1091	1209	1005	983	726	527	339	7874
Medford	0	0	78	372	678	871	918	697	642	432	242	78	5008
Pendleton	0	0	111	350	711	884	1017	773	617	396	205	63	5127
Portland	25	28	114	335	597	735	825	644	586	396	245	105	4635
Roseburg	22	16	105	329	567	713	766	608	570	405	267	123	4491
Salem	37	31	111	338	594	729	822	647	611	417	273	144	4754
Sexton Summit	81	81	171	443	666	874	958	809	818	609	465	279	6254
Pa. Allentown	0	0	90	353	693	1045	1116	1002	849	471	167	24	5810
Erie	0	25	102	391	714	1063	1169	1081	973	585	288	60	6451
Harrisburg	0	0	63	298	648	992	1045	907	766	396	124	12	5251

State & Station	Jul	Aug	Sep	Oct	Nov	Dec	Jan	Feb	Mar	Apr	May	Jun	Annual
Philadelphia	0	0	60	291	621	964	1014	890	744	390	115	12	5101
Pittsburgh	0	9	105	375	726	1063	1119	1002	874	480	195	39	5987
Reading	0	0	54	257	597	939	1001	885	735	372	105	0	4945
Scranton	0	19	132	434	762	1104	1156	1028	893	498	195	33	6254
Williamsport	0	9	111	375	717	1073	1122	1002	859	468	177	24	5934
R.I. Block Island	0	16	78	307	594	902	1020	955	877	612	344	99	5804
Providence	0	16	96	372	660	1023	1110	988	868	534	236	51	5954
S.C. Charleston	0	0	0	59	282	471	487	389	291	54	0	0	2033
Columbia	0	0	0	84	345	577	570	470	357	81	0	0	2484
Florence	0	0	0	78	315	552	552	459	347	84	0	0	2387
Greenville	0	0	0	112	387	636	648	535	434	120	12	0	2884
Spartanburg	0	0	15	130	417	667	663	560	453	144	25	0	3074
S.D. Huron	9	12	165	508	1014	1432	1628	1335	1125	600	288	87	8223
Rapid City	22	12	165	481	897	1172	1333	1145	1051	615	326	126	7345
Sioux Falls	19	25	168	462	972	1361	1544	1285	1082	573	270	78	7839
Tenn Bristol	0	0	51	236	573	828	828	700	598	261	68	0	4143
Chattanooga	0	0	18	143	468	698	722	577	453	150	25	0	3254
Knoxville	0	0	30	171	489	725	732	613	493	198	43	0	3494
Memphis	0	0	18	130	447	698	729	585	456	147	22	0	3232
Nashville	0	0	30	158	495	732	778	644	512	189	40	0	3578
OakRidge(co)	0	0	39	192	531	772	778	669	552	228	56	0	3817
Tex Abilene	0	0	0	99	366	586	642	470	347	114	0	0	2624
Amarillo	0	0	18	205	570	797	877	664	546	252	56	0	3985
Austin	0	0	0	31	225	388	468	325	223	51	0	0	1711
Brownsville	0	0	0	0	66	149	205	106	74	0	0	0	600
Corpus Christi	0	0	0	0	120	220	291	174	109	0	0	0	914
Dallas	0	0	0	62	321	524	601	440	319	90	6	0	2363
El Paso	0	0	0	84	414	648	685	445	319	105	0	0	2700
Ft. Worth	0	0	0	65	324	536	614	448	319	99	0	0	2405
Galveston	0	0	0	0	138	270	350	258	189	30	0	0	1235
Houston	0	0	0	6	183	307	384	288	192	36	0	0	1396
Laredo	0	0	0	0	105	217	267	134	74	0	0	0	797
Lubbock	0	0	18	174	513	744	800	613	484	201	31	0	3578
Midland	0	0	0	87	381	592	651	468	322	90	0	0	2591
Port Arthur	0	0	0	22	207	329	384	274	192	39	0	0	1447
San Angelo	0	0	0	68	318	536	567	412	288	66	0	0	2255
San Antonio	0	0	0	31	207	363	428	286	195	39	0	0	1549
Victoria	0	0	0	6	150	270	344	230	152	21	0	0	1173
Waco	0	0	0	43	270	456	536	389	270	66	0	0	2030
Wichita Falls	0	0	0	99	381	632	698	518	378	120	6	0	2832
Utah Milford	0	0	99	443	867	1141	1252	988	822	519	279	87	6497
Salt Lake City	0	0	81	419	849	1082	1172	910	763	459	233	84	6052
Wendover	0	0	48	372	822	1091	1178	902	729	408	177	51	5778
Vt Burlington	28	65	207	539	891	1349	1513	1333	1187	714	353	90	8269
Va Cape Henry	0	0	0	112	360	645	694	633	536	246	53	0	3279
Lynchburg	0	0	51	223	540	822	849	731	605	267	78	0	4166
Norfolk	0	0	0	136	408	698	738	655	533	216	37	0	3421
Richmond	0	0	36	214	495	784	815	703	546	219	53	0	3865
Roanoke	0	0	51	229	549	825	834	722	614	261	65	0	4150
Wash.Natl.AP	0	0	33	217	519	834	871	762	626	288	74	0	4224
Wash Olympia	68	71	198	422	636	753	834	675	645	450	308	177	5236
Seattle	50	47	129	329	543	657	738	599	577	396	242	117	4424
Seattle Boeing	34	40	147	384	624	763	831	655	408	411	242	99	4838
Seattle Tacoma	56	62	162	391	633	750	828	678	657	474	295	159	5145
Spokane	9	25	168	493	879	1082	1231	980	834	531	288	135	6655
Stampede Pass	273	291	393	701	1008	1178	1287	1075	1085	855	654	483	9283
Tatoosh Island	295	279	306	406	534	639	713	613	645	525	431	333	5719
Walla Walla	0	0	87	310	681	843	986	743	589	342	177	45	4805
Yakima	0	12	144	450	828	1039	1163	868	713	435	220	69	5941
W.Va Charleston	0	0	63	254	591	865	880	770	648	300	96	9	4476
Elkins	9	25	135	400	729	992	1008	896	791	444	198	48	5675
Huntington	0	0	63	257	585	856	880	764	636	294	99	12	4446
Parkersburg	0	0	60	264	606	905	942	826	691	339	115	6	4754
Wis Green Bay	28	50	174	484	924	1333	1494	1313	1141	654	335	99	8029
La Crosse	12	19	153	437	924	1339	1504	1277	1070	540	245	69	7589
Madison	25	40	174	474	930	1330	1473	1274	1113	618	310	102	7863
Milwaukee	43	47	174	471	876	1252	1376	1193	1054	642	372	135	7635
Wyo Casper	6	16	192	524	942	1169	1290	1084	1020	657	381	129	7410
Cheyenne	19	31	210	543	924	1101	1228	1056	1011	672	381	102	7278
Lander	6	19	204	555	1020	1299	1417	1145	1017	654	381	153	7870
Sherian	25	31	219	539	948	1200	1355	1154	1054	642	366	150	7683

APPENDIX E

HEAT DISTRIBUTION IN THE ATTACHED GREENHOUSE ON A CLEAR WINTER DAY

The following diagrams show the *approximate* heat distribution for two greenhouses. Example A is one attached to a frame structure, B is on a structure with a massive contiguous wall.

When you examine the two diagrams you see that the patterns are identical except for the F and G categories. The heat being conducted through the wall of a massive wall structure gives that example about a 20% edge in its capability to heat the attached structure over a similar greenhouse on a frame building.

The energy amount is for 1 sq. ft. of south glazing.

Example A

Greenhouse on a Frame Structure

2-3 ga. water per sq. ft. glazing is the thermal mass inside greenhouse

A	100%	(1600 BTUs)	Strikes Surface
B	20%	(320 BTUs)	Transmission Losses
C	5%	(80 BTUs)	To Photosynthesis
D	25%	(400 BTUs)	Evaporation, Conduction and Convection (infiltration) Losses
E	5%	(80 BTUs)	Foundation Losses
F	25%	(400 BTUs)	Thermal Mass inside greenhouse
G	5%	(80 BTUs)	Absorbed by north wall
H	15%	(240 BTUs)	Convection to home
	100%	1600 BTUs	

A

Figure 214

Example B

Greenhouse on a Massive Wall

Wall contains 5 gal. water per sq. ft. glazing or masonry equivalent

A	100%	(1600 BTUs)	Strike Surface
B	20%	(320 BTUs)	Transmission Losses
C	5%	(80 BTUs)	To Photosynthesis
D	25%	(400 BTUs)	Evaporation, Conduction and Convection (infiltration) Losses
E	5%	(80 BTUs)	Foundation Losses
F	5%	(80 BTUs)	Thermal mass inside greenhouse (floor and soil)
G	25%	(400 BTUs)	Absorbed by massive wall—for house and greenhouse
H	15%	(240 BTUs)	Convection to home
	100%	1600 BTUs	

B

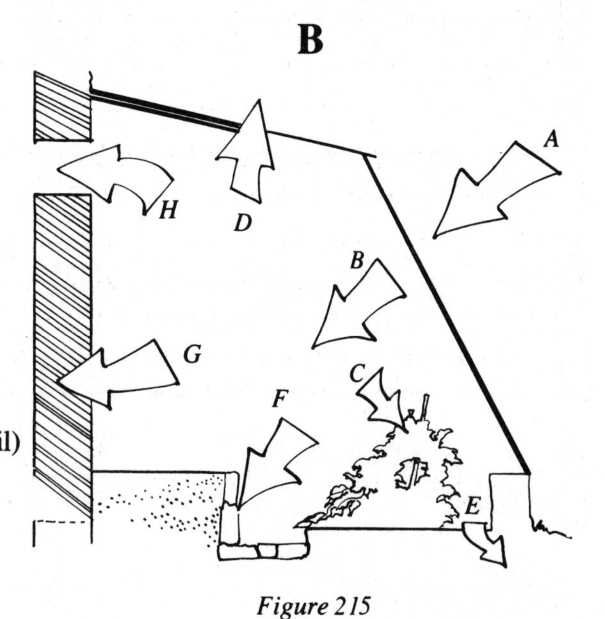

Figure 215

APPENDIX F

THERMAL STORAGE CAPACITY AND GREENHOUSE PERFORMANCE

The following formula will allow you to estimate the total heat storage of a greenhouse, as shown in Chapter VI.

By totaling the amount of thermal mass in the greenhouse we can get an estimation of how much total storage capacity we have.

Q_{st} = Total heat storage $Q_{st} = V \times d \times c \times \triangle t$
V = Volume in cubic feet (ft^3)
d = Density (lb/ft^3)
c = Specific heat constant
$\triangle t$ = Specific heat (BTU/lb°F)

We're going to make some assumptions here in $\triangle t$ based on field experience.

Material	*Placement*	*No. Units*	*V x*	(From chart p. 37) *d x*	*c x*	*$\triangle t$=*	*Q_{st}*
Water	55 gal. drums	6	330 gals.= 44 cubic ft.	62.5	1.00	20	55,000
Water	25 gal. drums	6	150 gals.= 20 cubic ft.	62.5	1.00	35	43,750
Stone	Planter 1 ft. wide x 8 ft. long x 3 ft. high	1	24 cubic ft.	165	.21	30	24,948
Brick	Floor 4'' x 160 sq. ft.	1	53 cubic ft.	120	.20	30	38,160
Earth	In Planters 12'' x 2 x 4	3	24 cubic ft.	95	.21	10	4,788

Total Thermal Storage in BTUs = 166, 646

This amount of storage would take the greenhouse through 2 days of heavy overcast with about 30°F rise over outdoor air temperatures.

APPENDIX G

HEAT CALCULATIONS FOR GREENHOUSES

A more conventional way (than my chart, Appendix A) of predicting the thermal performance of a solar greenhouse is found in the following article from *Alternate Sources of Energy Magazine #36* by Kolstad and Grimmer. Their method gives a wall credit to the attached greenhouse, but, it should be noted, it does not consider using night insulation.

Heat Losses

Heat losses in a tight (no air leaks) greenhouse will be due primarily to heat conduction to the outside. The simple physical law governing conductive losses states that the rate of heat loss through a surface is proportional to the temperature difference between the two sides of that surface:

$$Q = U \times A \times \triangle T$$

where Q is the amount of heat passing through the surface in a unit of time; $\triangle T$ the temperature difference across the surface; A the area of the surface and U the thermal conductance, a constant for the particular substance (of a given thickness). To give an example, if Q is measured in BTU/hour, a normal double glazed window has a thermal conductance U of approximately .69. If the area, A , is one square foot and $\triangle T$ is 30°F (outside averages 35°F, inside 65°F), then

$$Q = .69 \times 30 \times 1 = 20.7 \text{ BTU/hour.}$$

So, if we know the temperature difference between the inside and outside of the greenhouse and know the thermal characteristics (U) of the greenhouse material, we can find the daily heat loss. Thermal conductances of various substances are given in Table I. Experience with solar greenhouses in the sun and cold Southwest has lead to the conjecture that typical double glazed greenhouses maintain a 30°F temperature elevation over their surroundings. How accurate this may be in general is not known since it depends on many factors, particularly the weather. Nevertheless, let us use it as the temperature difference, $\triangle T$, in the equation above, recognizing that our results may not be too precise.

Table I:
Thermal Conductance
(BTU/hour/°F/ft^2)

(Derived from ASHRAE data)	U(=1/R)
1. Frame roof or wall	
With R11 insulation	0.07
With R22 insulation	0.04
With R33 insulation	0.03
2. Single glass	1.13
3. Double glass	
With small air space	0.69
With 1'' - 4'' air space	0.56
4. Triple glass	0.36
5. Solid wooden doors	0.49
6. Footing insulation (R5) (per linear foot)	0.4

Heat Availability

Suppose we now wish to build a greenhouse onto a particular dwelling. We can easily utilize the above discussion to obtain a rough estimate of how much useful heat we can expect to obtain from our greenhouse and consequently how much we can expect to save on our heating bills.

Step I: House Heating Needs

Find the amount of heat usually required by the dwelling on a monthly basis through a typical heating season (in, say, million of BTUs). This can be done by examining previous heating bills (remembering that furnaces burning oil or gas are usually only 60 percent efficient), or by doing a load analysis such as we discussed in the context of the greenhouse heat losses above. This need only be done for the actual heating season, say October-April. An important factor to include is that by placing a greenhouse against an outside wall, you are reducing heat losses through that wall. This credit should be subtracted from your real heating needs. The credit can be estimated by considering that the greenhouse reduces the temperature difference over the wall by 30°F. Thus the monthly savings is 30 days x U x Area x 24 hours x 30, where U is the thermal conductance of the house wall.

Step II: Greenhouse Gains

The monthly energy incident in sunlight must be corrected for a) the tilt and orientation of the greenhouse glazings and b) the transmission efficiency of the glazing. The solar flux on a horizontal surface ranges from 350-2000 BTU/day during the winter over the US. More precise data can and should be obtained for your paticular locale. This must then be adjusted for the orientation of the greenhouse glazings with respect to the horizontal. This information (solar flux and correction factors) is published in many of the good solar heating for the homeowner books (refer to the Bibliography). When the total incident light is determined based on total glazing area, a correction must be made for the losses incurred when the light passes through the glazing (and for coupling to the storage). A 50 percent efficiency factor is probably not a bad rule of thumb for double glazings.

Step III: Greenhouse Losses

Greenhouse losses will be assumed to be conductive losses. The total square footage of each of the major structural members of the greenhouse (glazing, roof, walls, etc.) exposed to the outside should be determined. Using the conductances (U) from Table I and a temperature difference, $\triangle T$ of 30°F, the total average heat loss in BTU/hour for the greenhouse can be obtained by summing, for all the components, the product of the appropriate U, 30°F and the area of the component. (In practice, it seems that for a floor, instead of area, it is better to use exposed perimeter). This gives a daily "load" which may be as much as 100,000 BTU/day or more.

Step IV: Net Available Heat

The last step is simple and consists of determining how much surplus heat (if any) will be available from the greenhouse to use in the home. This is done on a monthly basis. Available heat from the greenhouse will be assumed to be the monthly gain less the monthly loss. (Providing the difference is non-negative.) The house can be assumed to utilize this *up to* the monthly requirements of the house (from Step I). To give a rough idea of cost savings, each million BTU of heat saved by using the greenhouse will save from $3 to $15 (or more), depending on the fuel normally used for heating.

An Example

Suppose we live in Grand Junction, Colorado (39° North Latitude) and are contemplating building a greenhouse along the south side of our house. The greenhouse will be 20 feet long and will have 180 sq. ft. of double glazing facing south and tilted at 60° from the horizontal. The remainder of the greenhouse consists of 176 sq. ft. of wall and roof with 6 inches of fiberglass insulation (R22). The perimeter of the floor (except for the house/greenhouse wall) is 36 ft.

Step I: House Heating Needs

By examining our fuel oil bills for the past few seasons, we find average use of

Heating oil	OCT	NOV	DEC	JAN	FEB	MAR	APR
Used (gal)	27	63	85	99	90	63	36

Additionally, the greenhouse covers approximately 160 sq. ft. of house resulting in reduced losses through the house wall of 160 x 30 x 0.07 x 24 x 30 = 0.24 million BTUs/month (assume house wall has R11 insulation).

Using an average of 120,000 BTU/gal of fuel oil, we can translate the above oil use figures and the 0.24 million BTU saved per month figure into real heating needs:

	OCT	NOV	DEC	JAN	FEB	MAR	APR	Total
Heat Needs	1.9	4.5	6.1	7.1	6.5	4.5	2.6	33.2
Wall Credit	.24	.24	.24	.24	.24	.24	.24	1.7
Net Heating Req.	1.66	4.26	5.86	6.86	6.26	4.26	2.36	31.5

Real Heating Needs (Million BTU/month)

Step II: Greenhouse Gains

We have one glazing surface in our example, a southern tilted surface. From *The Solar Home Book* we find that for our location (approx. 40° North Latitude), the following average fluxes on the southern surface (BTU/month/ft^2):

	OCT	NOV	DEC	JAN	FEB	MAR	APR
Clear Day Flux	64294	57240	55676	60264	60928	67394	58680
Mean % Sun	70	60	60	50	60	60	60
Net Flux	45000	34350	33400	30100	36550	40450	35200

Grand Junction Solar Flux on Surface (BTU/month/ft^2)

Multiplying these numbers by our 50 percent transmission efficiency and the square footage of the glazing (180 ft^2) we obtain an estimate of total heat incoming to the greenhouse (in millions of BTU/month).

	OCT	NOV	DEC	JAN	FEB	MAR	APR
Per ft^2	.0225	.0172	.0167	.0151	.0183	.0202	.0176
Total	4.05	3.09	3.01	2.71	3.29	3.64	3.17

Step III: Greenhouse Losses

From Table I, we can obtain thermal conductances for the three basic greenhouse structural components and estimate daily heat losses:

Daily Loss for Glazings:	.56 x 30 x 180 x 24	72576 BTU
Daily Loss for Roof and Walls:	.04 x 30 x 176 x 24	5067 BTU
Daily Loss Through Floor:	.4 x 30 x 38 x 24	10944 BTU
		88587 BTU/day

for a total monthly loss of 2.66 million BTUs.

Step IV: Net Available Heat

We summarize the previous steps below and obtain net monthly available heat (in millions of BTU):

	OCT	NOV	DEC	JAN	FEB	MAR	APR
Greenhouse Gains	4.05	3.09	3.01	2.71	3.29	3.64	3.17
Greenhouse Losses	2.66	2.66	2.66	2.66	2.66	2.66	2.66
Net from Greenhouse	1.39	0.43	0.35	0.05	0.63	0.98	0.51
House Requirements	1.9	4.5	6.1	7.1	6.5	4.5	2.6
Wall Credit	.24	.24	.24	.24	.24	.24	.24
Net to House	1.63	0.67	0.59	0.29	0.87	1.22	0.75

This indicates 6.02 million BTU were supplied to the house, 18 percent of total house requirements.

Our purpose here has been to outline some ot the major considerations in a performance analysis of a greenhouse as a solar heating system. One should not be mislead that the add-on greenhouse will result in tremendous monetary or energy savings to the user. In order to save money on one's heating bill, a considerable investment in time and money must be made. What makes the attached solar greenhouse so attractive is its triple function—a solar collector for home heating, a pleasant solar room in winter, and a source of food and plants throughout the year.

Taken together, the solar greenhouse becomes a very attractive option.

Charles Kolstad is an applied mathematician/economist who has been involved with the economic analysis of energy systems since 1974. Derrick Grimmer is a natural philosopher who has done research in low-temperature physics and solar energy. Both authors have been active in appropriate technology for a number of years and are directors of Appropriate Technology Research, a non-profit coalition of scientists in Santa Fe, New Mexico (P.O. Box 5852).

APPENDIX H

When planning openings between the house and greenhouse for winter heating we are trying to achieve an airflow that will: 1) put usable heat into the home (+75°F) and, 2) leave the greenhouse in an acceptable temperature range (70-80°F). A rate has been established that does both of these things. The rule of thumb is to move *4-6* cubic feet per minute of air *per* each square foot of south glazing.

The following formula can be used to size vents, windows and doors for passive air circulation.

$$16.6 \times a \sqrt{\Delta T \times h}$$

Here's what the symbols stand for:

a= area in sq. ft. of the *smallest* vent in a high/low series.
16.6 = a factor accounting for passive air movement.
ΔT = the temperature *difference* between the high and low vent in Farenheit. This is often an assumption.
h = the average height *difference* between high and low openings in feet. You can measure or estimate from center to center.

Let's work the formula with an example. We will plug-in some temperatures that are derived from experience with attached greenhouses. For the vents we will try a window and door combination. The door is at the floor level of both the house and greenhouse. The window is located slightly higher than the door. The door can act as the lower vent but only the bottom two feet will serve this purpose. Both the window and door have about 6 sq. ft. of effective vent area. Temperatures at the bottom of the door, coming from the floor of the home, are 60°F. Temperatures at the window level will be around 85° on a clear winter day. Measuring, we find that there is 6 feet of vertical distance between the lower part of the door and the center of the window. So:

a = 6 sq. ft.
ΔT = 25°
h = 6 ft.

$$(16.6)(6) \quad \sqrt{(25)(6)} = 12.19.84\text{cfm}$$

1220 cubic feet per minute will move through the openings. If the greenhouse has 200 sq. ft. of south glazing, the size of the openings is just about right. (200 x 6 = 1200).

The equation can also be used to approximate the size and distance between the vents used for summer exhaust. Looking at the formula, it becomes obvious that the *most important factor* is the square footage of the inlet and exhaust vents. Next in priority is the distance between vents. (The Δt is something you have to live with in that the working greenhouse has certain temperature limits that can't be exceeded.)

Here's another example. You have a greenhouse that is to be used throughout the summer. You want passive venting only. You desire to have summer ventilation at the rate of *one total greenhouse air exchange per minute* (a standard quantity in cooling greenhouses). If you also have cold winters, a long series of smaller vents along the bottom and top of the greenhouse isn't recommended because of greater infiltration losses along the perimeters of the vents. Outside highs peak at about 85° and the apex temperatures of the greenhouse can be 100°. Your smallest vent size is 6 sq. ft. The vertical distance between the vents is 8 feet. The volume of the greenhouse is 1,600 cubic feet (16' L x 10H x 10'D) and that's what has to be moved on a per minute basis.

a = 6 sq. ft.
ΔT = 15°
h = 8

$$(16.6)6 \sqrt{(15)(8)} = 1091.06 \text{ cfm}$$

1091 cubic feet per minute. You're far short of the goal. It would appear that we must increase the size of the vents in order to get adequate air flow through them. That's correct and is probably the easiest way to solve the problem. But there's another passive solution, the *thermal chimney*. Suppose you were to build a plywood stack over the upper vent to a height of 10 feet. This would increase your average distance between openings of 18 feet. Furthermore, what if you painted it black? It's reasonable to assume that it would raise temperatures inside the stack in summer to 110°. Then we would have:

$a = 6$ sq. ft.
$\Delta T = 25^{\circ}$
$h = 18$ ft.

$$(16.6)6\sqrt{(25)(18)} = 2112.83\text{cfm}$$

2113 cubic feet per minute or even more than the amount we needed. That's fine, we'll welcome the additional circulation in summer.

This simple example should point out what happens when either the distance between vents and/or the temperature difference is increased. The equation can be used to approximate air flow in any application where the heights, temperatures, and vent sizes can be estimated (trombe wall, attic ventilation, etc.).

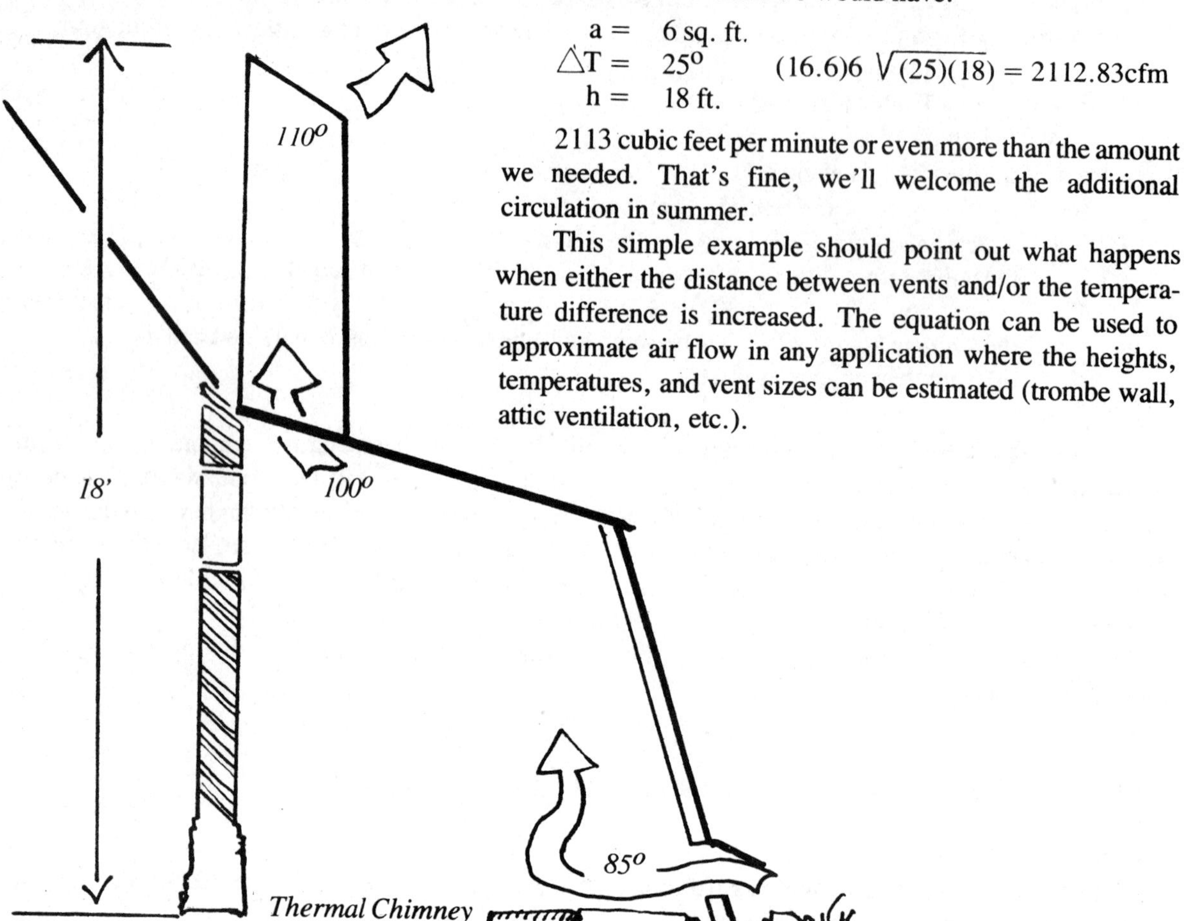

Figure 216

APPENDIX I

PLANTING CHART

Plant	Varieties	Location	Time of Year to Plant	Special Instructions
Tomato	Earliana Marglobe Michigan Ohio Early Girl Sweet 100 Any small variety patio, cherry, pear	Front of greenhouse in spring. On back wall in winter.	Early spring Mid-August	Need full photoperiod. Pollinate by lightly tapping open blossoms or shake plant vigorously. Train plants up strings. Trim foliage severely when infested with insects and in the fall to prevent shading of the greenhouse. Do not cut top growth until you are ready for plant to stop growing. Tomatoes are perennials and will produce for a long time. Pull suckers (found in crotch of limbs) off. They can be rooted: start in sand or vermiculite.
Cucumber	Toska Gourmet	Clear side of greenhouse in Spring. Back wall in winter. Can stand some shading.	Early Spring Mid-August	Need full photoperiod. Pollinate with a small brush or let the bees in. Pull off first several feet of blossoms for better fruit set. Train on string or twine. Can be trained to climb all over the sides and roof of greenhouse.
Peppers	Red Chile Any green, bell wax	Full light area	Early Spring Mid-August	Adapt well to small container or beds. Pollinate with small brush. Be careful not to over-water. Chile Peppers do not seem to get as hot as they do outdoors; still delicious, however.
Melons and Squash	Watermelon Cantaloupe Pumpkin Honeydew Crookneck Zucchini Acorn	Need light and lots of room. Front of greenhouse.	Early Spring	Trim vegetation. Grow out the vents—summer. Will cross pollinate: try to separate varieties by distance.
Leafy Greens	Leaf Lettuce Endive Kale Spinach Mustard Greens Cress Chard Collards Chicory Celtuce	Medium light. Cool.	Any time—makes the most sense in late Fall, Winter, early Spring.	Dependable winter producers. Head lettuce does not head well in greenhouse. Plant densely, thin as you eat. Leafy greens will grow in pots, on vertical surfaces, almost anywhere. Plant under trimmed tomatoes or cucumbers. Can be cut many times while growing. Will go to seed if temperatures get too hot.
Carrots	Smaller varieties	Sunny	Fall Winter Early Spring	Plant thickly and thin out. Slow maturers. Interplant with tomatoes.

Plant	Varieties	Location	Time of Year to Plant	Special Instructions
Beets Turnips	Any	Medium Light	Fall Winter	Do well in shallow boxes. Plant thickly, but thin to allow root to become large. Foliage when small makes good edibles.
Radishes	Any	Any place	Anytime	Don't plant more than you can eat. Excellent indicator of soil viability: should sprout in 3-5 days.
Broccoli Cauliflower Cabbage Brussel Sprouts	Calabrese, Italian Snowball Golden acre, Chinese Jade Cross	Medium light. Cool	Late Summer—Early Fall for Winter heading	Spatially consuming crops. Do well in pots. Transplant well into garden.
Beans	Pole Burpee Golden Blue Lake	Medium light. Up walls.	Early Spring. Late Summer.	Great on north wall of greenhouse. Train on trellises. Climbers can be used on greenhouse exterior for shade (Red Pole Beans).
Eggplant	Black Beauty Early Beauty	Light area.	Early Spring. through Summer.	Pollinate with small brush or fingertip. Transplant from garden back to greenhouse in fall.
Peas	Burpeeana Early Blue Bantam Snow Sugar	Shady Areas. Cool.	Fall Winter Early Spring	Used to replenish nitrogen in soil. Four poles in corners, string lattices across.
Onions Scallions Garlic	Any	Medium light	Fall Winter	Keep soil moist. Fresh tops are great in salads, trim regularly. start seeds in greenhouse for garden sets. Garlic is an insect fighter, either growing or ground into water solution.
Strawberries	Everbearing	Shade. Under tables.	Fall.	Like more water than vegetables.

Herbs — We have had tremendous success with all we have tried. They grow in any location, but prefer medium light. Most will do well in cold weather.

Anise Basil	Borage Chives	Caraway Chervil	Coriander Tarragon	Oregano Mint (all kinds)	Parsley Sage	Thyme Sweet Marjoram	Dill

Transplants Out

Best for transplant from greenhouse to garden. Don't be deceived. Start 6–8 weeks before anticipated last frost. They grow fast. Try not to disturb root system any more than possible: melon and cucumber roots are easily damaged. Harden off by exposing to outdoor temperatures two weeks before transplanting. Grow in small containers when you can: we have used Jiffy pots (these need to cut before being put into the ground; they do not disintegrate quickly enough and the plant can become root bound), expanding peat pellets, styrofoam cups, milk cartons, MacDonalds' quarter pounder containers, tin cans. If possible, transplant on a cloudy day, shade and water well after transplanting. Do not forget that all of these seedlings plus flower and herb starts are big sellers in the Spring. A small (160 square foot) solar greenhouse I built in Idaho sold $273.00 worth of seedlings its first Spring. That was over half the cost of the greenhouse.

Tomatoes Peppers Melons Broccoli Eggplant Sunflowers Celery Cucumbers
Squash Corn

Transplants In

If you are careful, you can transplant healthy garden crops back into the greenhouse in the fall. Be sure to check for insects and disease first. Get all the root system you can, shade and water them in their new home.

Tomatoes Peppers Eggplant Melons Onions Broccoli Petunias Pansies
Marigolds Nasturtiums Geraniums

Daily Temperature Chart for Solar Greenhouse

Month of: ____________________

Date	Temperature				WEATHER CONDITIONS (check one)						Comments
	Inside		Outside								
	Hi	Low	Hi	Low	Clear	Partly Overcast	Overcast	Rain	Snow	Wind	

From Solar Sustenance Team
Santa Fe, NM 87501

Variety	Date Planted	Date Germinated	Dates of Harvest	Bed Location

From Solar Sustenance Team
Santa Fe, NM 87501

Time Spent in Solar Greenhouse

Month of: ____________________

Date	Amount of Time	What Done (planting, weeding, fertilizing, sitting & enjoying)	Comments

From Solar Sustenance Team
Santa Fe, NM 87501

BIBLIOGRAPHY

Appropriate Technology

Small is Beautiful, E.F. Schumacher, Harper and Row, 1973
Energy Primer, Portola Institute, Dell Publishing Co., 1978
Radical Agriculture, Richard Merrill, editor, Harper and Row, 1976
A Landscape for Humans, Peter van Dresser, Biotechnic Press, 1975
Architecture for the Poor, Hassan Fathy, The University of Chicago Press, 1969

General Solar

Direct Use of the Sun's Energy, Farrington Daniels, Ballantine Books, 1964
The Coming Age of Solar Energy, D.S. Halacy, Harper and Row, 1973
Solar Energy and Shelter Design, Bruce Anderson, Total Environmental Action, 1973
The Solar Home Book, Bruce Anderson, Cheshire Books, 1976
Sunspots, Steve Baer, Biotechnic Press, 1975
Sunset Homeowner's Guide to Solar Heating, Lane Publishing Co., 1978
Natural Solar Architecture, a passive primer, David Wright, Van Nostrand Reinhold Co., 1978
Other Homes and Garbage, Leckie, Masters, Whitehouse and Young, Sierra Club, 1975
Home Grown Sundwellings, Peter van Dresser, The Lightning Tree, 1977
Solar Heated Buildings of North America, William Shurcliff, Brick House Publishing Co., 1978
New Inventions in Low Cost Solar Heating, William A. Shurcliff, Brick House Publishing Co., 1979
The Passive Solar Energy Book, Ed Mazria, Rodale Press, 1979
Solar for Your Present Home, Berkeley Solar Group and Lynn Nelson, California Energy Commission, 1977
Handbook of Fundamentals, ASHRAE, 1972 (especially Chapter 59 for Solar Data)

Solar Greenhouse Design and Construction

The Unheated Greenhouse, D. Goold-Adams, Transatlantic Arts, Inc., 1955
The Complete Greenhouse Book, Peter Clegg and Derry Watkins, Garden Way Publishing, 1978
The Solar Greenhouse Book, James McCullagh, editor, Rodale Press, 1978
Your Homemade Greenhouse and How to Build It, Jack Kramer, Cornerstone, 1975
An Attached Solar Greenhouse, W.F. and Susan Yanda, The Lightning Tree, 1976 (in Spanish and English)
Vocational Region 10 Solar Greenhouse, Maine Audubon Society, 1978
A Solar Greenhouse Guide for the Northwest, Ecotope Group, 1978

Construction

Building Construction Illustrated, Francis Ching, Van Nostrand Reinhold, 1976
The Owner Builder and the Code: Politics of Building Your Own Home, Ken Kern, Ted Kogon and Rob Thalon, Scribner's, 1976
The Owner Built Home, Ken Kern, Scribner's, 1975
Low-Cost Energy Efficient Shelter, Eugene Eccli, Rodale Press, 1971
Dwelling House Construction, Albert Dietz, The M.I.T. Press, 1971
Do-It-Yourself Housebuilding: Step by Step, MacMillan, 1973
Basic Construction Techniques for Houses and Small Buildings Simply Explained, Bureau of Naval Personnel, Dover
From the Ground Up, Cole and Wing, Little Brown and Co., 1976
From the Walls In, Charles Wing, Little Brown and Co., 1979

Periodicals There are so many good periodicals out now. If you look, you will probably find an excellent newsletter or bulletin published in your locale. Listed below are a few nationally distributed publications that have emphasized passive solar and solar greenhouse work.

Southwest Bulletin, New Mexico Solar Energy Association, P.O Box 2004, Santa Fe, NM 87501
Solar Greenhouse Digest, P.O. Box 2621, Flagstaff, AZ 86003
Solar Age Magazine, Solar Vision, Inc., Harrisville, NH
Solar Engineering Magazine, Dallas, TX
RAIN, Portland, OR

Popular Science and *Popular Mechanics* are both giving increasing emphasis to solar and passive applications.

Alternate Sources of Energy, Milaka, MINN
Mother Earth News, Hendersonville, NC
Organic Gardening and Farming, Emmaus, PA
New Roots, Amherst, MA

Conference Proceedings are your best source of technical information. As a matter of fact, if you are only after the technical information, skip going to the conferences and just order the proceedings.

Conference on Energy Conserving Solar Heated Greenhouses, Marlboro College, Marlboro, Vermont, 1978, $9.00
Great Lakes Solar Greenhouse Conference, Chippewa Nature Center, 400 S. Badour Rd., Rt. #9, Midland, MI 48640, 1978, $7.00
Conference on Solar Energy for Heating Greenhouses and Greenhouse-Residential Combinations, Sponsored by Ohio Agricultural Research Development Center and ERDA, 1977
Third Annual Conference: Solar Energy for Heating Greenhouses and Greenhouse-Residential Combinations, U.S.D.A./D.O.E., 1978
Passive Solar Heating and Cooling, First Passive Conference, National Technical Information Service, U.S. Dept. of Commerce, Springfield, VA 22161, 1976, $10.50
Passive Solar, State of the Art. Second Passive Conference. Mid-Atlantic Solar Energy Association, 2233 Gray's Ferry Ave., Philadelphia, PA 19146, 1978, $20.00
Third National Passive Solar Conference. For information write: AS/ISES, P.O. Box 1416, Killeen, TX 76541
Fourth National Passive Solar Conference. For information write: AS/ISES, P.O. Box 1416, Killeen, TX 76541.

Gardening

How to Grow More Vegetables Than You Ever Thought Possible on Less Than You Can Imagine, John Jeavons, Ecology Action of the Mid Peninsula, 1974
Gardening for People (Who Think They Don't Know How), Doug Moon, John Muir Publications, 1975
The Postage Stamp Garden Book, Duane Newcomb, Hawthorn, 1975
The City People's Book to Raising Food, Helga and William Olkowski, Rodale Press, 1975
How to Grow Vegetables and Fruit by the Organic Method, J.I. Rodale and staff, Rodale Press, 1974
The Encyclopedia of Organic Gardening, J.I. Rodale and staff, Rodale Press, 1975
The Basic Book of Organic Gardening, Robert Rodale, editor, Rodale Press, 1975
How to Have a Green Thumb Without an Aching Back, Ruth Stout, Cornerstone, 1974
The Gardener's Catalog, produced by Harry Rottenberg, William Morrow and Co., 1974
Food Gardens: Indoors, Outdoors and Under Glass, produced by Harry Rottenberg, William Morrow and Co., 1975

Crockett's Victory Garden, James Crockett, Little Brown and Co., 1977
Crockett's Indoor Gaden, James Crockett, Little Brown and Co., 1978
The Green Pages, Maggie Oster, editor, Tree Communications, Inc., 1977
Organic Gardening Under Glass, George and Katy Abraham, Rodale Press, 1975
Hydroponic Gardening, Raymond Bridwell, Woodbridge Press, 1972
The Survival Greenhouse, James B. DeKorne, The Walden Foundation, 1975
Beginner's Guide to Hydroponics, James S. Douglas, Drake Publishers, Inc., 1972
Gardening Indoors Under Lights, F.H. Franz and J.L. Kranz, The Viking Press, 1957
The Solar Greenhouse Book, James McCullagh, editor, Rodale Press, 1978
Winter Flowers in the Greenhouse and Sun Heated Pit, K.S. Taylor and E.W. Gregg, Charles Scribner and Sons, 1969
The Facts of Light, Ortho Book Series, 1975
The Complete Book of Greenhouse Gardening, Ian Walls, Quadrangle, 1975
Grow Fruit in Your Greenhouse, George E. Whitehead, Faber and Faber, London, 1970
What Every Gardener Should Know About Earthworms, published by Garden Way Research
A to Z of Greenhouse Plants: A Guide for Beginners, Paul Langfield, Max Parrish and Co., 1964
Weeds: Guardians of the Soil, Joseph A. Cocannouer, Ecology Action
Companion Plants and How to Use Them, Helen Philbrick and Richard Gregg, Ecology Action
Gardening Without Poisons, Beatrice Trum Hunter, Houghton Mifflin, 1972
Common Sense Pest Control, Helga Olkowski, Consumer's Cooperative of Berkeley, Inc., 1971
Organic Plant Protection, Rodale Press, 1972
The Gardener's Bug Book, Cynthia Westcott, Doubleday and Co., Inc., 1973
Rodale Herb Book, Rodale Press
A Modern Herbal, M.A. Grieve, Dover Publications, 1971
Herbs and Things, Jeanne Rose, Grosset and Dunlap, 1969
Composting: A Study of the Process and Its Principles, Clarence G. Golueke, Rodale Press, 1972

Seed Catalogs There are many more seed companies than we could ever list here. Included are some little known suppliers with unusual specialties and some well-known seed houses for their more complete selection. In ordering catalogs look for seed companies familiar with your climatic conditions that cater to home gardeners.

Abundant Life Seeds, Box 30018, Seattle, WA 98103
Small seed exchange for residents of the Pacific Northwest and California only. Beautiful origami catalog folded on one page for 50¢.

Burpee Seed Co., Clinton, IA 52732
Large well-known company with wide selection of most vegetables and flowers.

DiGiorgi Co., Council Bluffs, Iowa 51501
Forage crops, old-fashioned lettuce and other vegetables, open-pollinated corn.

Gurney's, Yankton, SD 57078
Unusual vegetables. Cold-weather vegetables and fruit trees.

Hart Seed Co., Wethersfield, CT 06109
Largest selection of old-fashioned and non-hybrid vegetables. Many hard-to-find varieties available on request.

J.L. Hudson Seed Co., P.O. Box 1058, Redwood City, CA 94064
One of the world's largest selection of flower and herb seeds. 50¢ for catalog.

Johnny's Selected Seeds, Albion, ME 04910
Small seed company with integrity. Carries native american crops, select oriental vegetables, grains, and short-maturing soybeans. Catalog: 50¢

Lanark County North American Medicinal and Culinary Herbs, RR 2, Amonte, Ontario
Canadian seed house, good for cold-weather areas. Ask for free organically-grown vegetable seed price list.
Meadowbrook Herb Garden, Rt. 138, Wyoming, RI 02898
Bio-Dynamically grown spices, herbs, teas and herb seeds.
Nichols Garden Nursery, 1190 North Pacific Hwy., Albany, OR 97321
Unusual specialities: elephant garlic, luffa sponge, winemaking supplies, herbs.
Park Seed Co., Greenwood, SC 29647
The best selection of flowers. Gorgeous, full-color catalog available free.
Redwood City Seed Co., P.O. Box 361, Redwood City, CA 94061
Basic selection of non-hybrid, untreated vegetable and herb seeds. Expert on locating various tree seeds, including redwoods. Guide on saving seed for 75¢. Catalog costs 25¢.
R.H. Shumway, Rockford, IL 61101
Good selection of grains, fodders and cover crops.
Stark Brothers, Louisiana, MO 63353
Specializes in fruit trees, especially dwarfs and semi-dwarfs. Carries many developed by Luther Burbank.
Stokes Seeds, Box 548, Buffalo, NY 14240
Carries excellent varieties of many vegetables, especially carrots.
Sutton Seeds, London Road, Earley, Reading, Berkshire, England RG6 1AB
For gourmet gardeners. Excellent, tasty varieties, hot-house vegetables.
Vilmorin-Andrieux, 4 Quai de la Megisserie, 75001 Paris, France
Old, respected seed house specializing in high-quality gourmet vegetables. Catalog in French.
Arthur Yates & Co., P.O. Box 72, Revesvy 2212, New South Wales, Australia
Specializes in tropical varieties suitable for the southern hemisphere. International seed catalog free.

Suppliers and Catalogs

Solar Usage Now, Box 306, Bascom, Ohio 44809
A jam packed catalog of solar supplies. Ships mail order.
Solarware, Pawlet, Vermont 05761
A new passive hardware catalog.
A to Z Solar Products, 200 East 26th Street, Minneapolis, MINN 55404
A good catalog and source book.
Solar Age Catalog, Solar Vision, Inc. Manchester, New Hampshire
Good access to products, ideas and people.
Kalwall Catalog, 88 Pine Street, Manchester, New Hampshire 03103

Greenhouse Supplies The following companies have extensive catalogs carrying all types of greenhouse products and accessories. They deal primarily with large commercial operations.

National Greenhouse Company, Pana, Illinois 62557
Stuppy, Inc., Greenhouse Supply Division, P.O. Box 12456, Kansas City, MO 64416
Texas Greenhouse Company, 2717 St. Louis Ave., Fort Worth, TX 76110
George J. Ball, Inc., Box 335, West Chicago, IL 60185

Glazing Suppliers

ASG Industries, Inc., P.O. Box 292, Kinsport, TN 37662
Glass glazing information.
CY-RO Industries, Berden Ave., Wayne, NJ 07470
"Acrylite SDP" double skinned acrylic sheet.

Brother Sun, Rte. 6, Box 10A, Santa Fe, NM 87501
Acrylite SDP, Lascolite, Lexan glazings and greenhouse components.
Filon Division of Vistron Corporation, 12333 Van Ness Ave., Hawthorn, CA 90250
Fiberglass glazing.
Kalwall Corporation, 88 Pine St., Manchester, NH 03103
Fiberglass glazing, fiberglass water heat storage tubes.

People Resource

Rainbook, Schocken Books, New York, 1977
Citizens Energy Directory, 2nd edition, 1110 6th St., NW, Washington, DC 20001

Hand Held Calculator Programs for Solar Greenhouse Performance

These programs can quite accurately predict the performance of a greenhouse on a seasonal or daily basis. They are most appropriate for architects and engineers.

TEANET Total Environmental Action, Inc., Church Hill, Harrisville, NH 03456
$95.00 from TEA
PEGFLOAT Princeton Energy Group, 729 Alexander Road, Princeton, NJ 08540
$75.00 combined with PEGFIX
PEGFIX Princeton Energy Group, 729 Alexander Road, Princeton, NJ 08540
$75.00 combined with PEGFLOAT
SUN-PULSE II Joint Venture, 138 Mt. Auburn St., Cambridge, MA 02138
$100.00 from above or McGraw Hill Book Co.

Audio Visual Materials

Passive Solar Energy Slide Set, 6 slide sets. Set E is *Attached Greenhouses* prepared by the New Mexico Solar Energy Association, P.O. Box 2004, Santa Fe, NM 87501
Solar Greenhouse Slide Series, 7 slide sets covering all aspects of the solar greenhouse prepared by The Solar Sustenance Team, Rte 1, Box 107AA, Santa Fe, NM 87501
Build Your Own Greenhouse-Solar Style, 16mm color/sound 21-minute film, available in French, Spanish and English. Danamar Film Production, 275 Kilby, Los Alamos, NM 87544

Regional Contacts This list, arranged alphabetically by state, is constantly changing and expanding. These groups, many of whom run greenhouse construction workshops, can put you in contact with the right people in your area. If writing for information, please enclose a SASE.

Solar Greenhouse Employment Project, Box 435, Tuscaloosa, ALA 35401
Northern Arizona Council of Governments, P.O. Box 57, Flagstaff, AZ 86002
The Ozark Institute, Box 549, Eureka Springs, AK 72632
Habitat Center, 162 Christian Drive, Pacheco, CA 94533
SUNRAE, 257 Santa Monica Way, Santa Barbara, CA 93109
Farralones Institute, 18290 Coleman Valley Road, Occidental, CA 95465
Grand Junction Public Energy Office, 250 North 5th, Grand Junction, CO 81501
San Luis Valley Solar Energy Association, P.O. Box 1284, Alamosa, CO 81101
Domestic Technology Institute, 12520 W. Cedar Drive, Lakewood, CO 80228
Regional Rehabilitation Institute, 157 Church Street, New Haven, CT 06510
Institute for Local Self Reliance, 1717 18th St., NW, Washington, DC 20009
Design Alternatives, 1312 18th St. NW, Washington, DC 20001
Citizen's Energy Project, 1110 6th St. NW, Washington, DC 20001

Solar Lobby, 1001 Connecticut Ave NW, Washington, DC 20036
Claude Terry and Associates, 929 Euclid Ave., NE #1, Atlanta, GA 30307
Center for Neighborhood Technology, 570 West Randolph St., Chicago, IL 60606
Frankfort Public Library, Frankfort, IN
Save the Children Federation, Berea, KY
Buck Neelis, 242 Grand, Lafayette, LA 70503
Maine Audubon Society, Gilsland Farm, 118 Old Route 1, Falmouth, ME 04105
New Alchemists, Box 432, Woods Hole, MA 02543
Massachusetts Energy Office, Division of Conservation and Solar, 73 Tremont St., Room 849, Boston, MA 02108
Northeast Solar Energy Center, 70 Memorial Drive, Cambridge, MA 02142
Chippewa Nature Center, 400 S. Badour Road, Rt. #9, Midland, MI 48640
Ingram County Energy Office, Cooperative Extension Service, 121 E. Maple, Mason, MI 48854
Center for Local Self Reliance, 3302 Chicago Avenue South, Minneapolis MINN 55407
Solar Greenhome Associaion, 34 N. Gore, Webster Grove, MO 63119
Montana Sunteam, P.O. Box 216, Circle, Montana 59215
Total Environmental Action, Church Hill, Harrisville, NH 03540
New Mexico Solar Energy Association, P.O. Box 2004, Santa Fe, NM 87501
Solar Sustenance Team, Rte, 1, Box 107AA, Santa Fe, NM 87501
New Mexico Energy Extension Service, P.O. Box 00, Santa Fe, NM 87501
Sunspace, Inc., P.O. Box 1792, Ada, OK 74820
Long Branch Land Association, Rte. 2, Box 132, Leicaster, NC 28748
Portland Sun, 1815 South East Main Street #4, Portland, OR 97214
Office of Community Energy, P.O. Box 156, Harrisburg, PA 17120
Governor's Energy Office, 80 Dean Street, Providence, RI
Tennessee Environmental Council, P.O. Box 1422, Nashville, TN 37202
Solar Applications Branch, Tennessee Valley Authority, 426 United Bank Building, Chattanooga, TN 37401
San Marcus Community Center, San Marcus, TX
Office of the City Manager, Crystal City, P.O. Drawer 550, Crystal City, TX 78839
Utah Solar Energy Association, 1159 East Stratford Avenue, Salt Lake City, Utah 84106
New England Solar Energy Association, P.O. Box 541, Brattleboro, VT 05301
Ecotope Group, 2322 East Madison, Seattle, WA 98112
Washington Energy Extension Services, Room 312, Smith Tower, Seattle, WA 98104
West Central Wisconsin Community Action Agency, 525 Second Street, Glenwood City, Wisconsin 54103
Community Relations, Soc. Dev. Com., 161 W. Wisconsin Ave., Milwaukee, WI 53202
Gale Harms, Wyoming Energy Extension Services, 521 Park, Thermopolis, WY 82443

All Local, State, and Regional Solar Energy Associations are very good sources of information. Look for them in the telephone book.

INDEX

- E -

- F -

- G -

- H -

- I -

- J -

- K -

- L -

- M -

- N -

- O -

- T -

- U -

- V -

- W -